Vincent van Gogh: *Self-Portrait.* 1889, St. Rémy, oil on canvas, 65 × 54 cm (25.6 × 21.3″) Paris, Musée d'Orsay, © Photo R.M.N.

Vincent van Gogh:
Chemicals, Crises, and Creativity

Wilfred Niels Arnold

Birkhäuser
Boston · Basel · Berlin

Library of Congress Cataloging-in-Publication Data

Arnold, Wilfred Niels, 1936–
Vincent van Gogh : chemicals, crises, and creativity /
Wilfred Niels Arnold.
 Includes bibliographical references and index.
 ISBN-13:978-1-4612-7742-2
 1. Gogh, Vincent van, 18531890 – Mental health. 2. Gogh,
Vincent van, 1853–1890 – Psychology. I. Title.
ND653.G7A85 1992
759.9492–dc20

Die Deutsche Bibliothek – CIP-Einheitsaufnahme

Arnold, Wilfred Niels:
Vincent van Gogh : chemicals, crises, and creativity / Wilfred Niels
Arnold. – Basel ; Boston ; Berlin : Birkhäuser, 1992
 ISBN-13:978-1-4612-7742-2   e-ISBN-13:978-1-4612-2976-6
 DOI:10.1007/978-1-4612-2976-6

Jacket design: Justin Messmer, Basel

9 8 7 6 5 4 3 2 1

ISBN-13:978-1-4612-7742-2

To my parents, Doris and Curt Arnold

Contents

Preface

As a five year old I encountered a picture of a young man in a rakish hat and a yellow coat, on the wall of a large classroom. There was something instantly intriguing about the image, but it was also puzzling because it represented neither politician nor prince, the usual fare for Australian school decorations. I was eventually told that this was a reproduction of a painting, the artist was Vincent van Gogh, and that the subject was some young Frenchman. On special days we assembled in that room and during the next several years I found myself gazing beyond visiting speakers at the fellow in the yellow jacket. It was almost another fifty years before I felt properly conversant with the portrait and realized that van Gogh's subject, Armand Roulin, was seventeen at the time of the original painting and had died at seventy-four during my schoolboy contemplations.

In the interim my enjoyment of the works of the Impressionists and Post Impressionists had grown and I occasionally ran into the name of Dr. Gachet, Vincent's last attending physician, in books and catalog essays. The doctor was my entrée to the overlapping charms of medical and art histories. In 1987 I had the good fortune to participate as a biochemist in the centenary celebration of the Pasteur Institut in Paris. On the weekend I took the train to Auvers-sur-Oise and visited the graves of Vincent and Theo van Gogh, and the home of Dr. Gachet. The present owner of the house, Colonel Vandenbroucke, pointed out a thuja tree that had previously decorated Vincent's first grave-site, and I made a surrealistic connection with thujone, the toxic principle of absinthe. I started brushing up on terpenes and related chemicals.

I read the correspondence of van Gogh again, discovered some additional chemical connections, and started synthesizing working hypotheses that might explain Vincent's underlying illness, the importance of exacerbation factors, and the occasional high yellow palette, all in the context of the nineteenth century and the artist's life style. Some journal papers followed, and I was encouraged to undertake a broader study.

The letters of Vincent van Gogh are a splendid record of his life, work, and philosophy. They have provided the primary source and substance of numerous studies, particularly by art historians and psychiatrists, who have delved into the writings of the artist from their own special points of view. Their products naturally range from pedantic essays on letter dating, through detailed analyses of personal interactions, to dissertations on mental illness. Especially with regard to psychological analyses there has been no shortage of speculation. But the more medical, nutritional, and environmental aspects of Vincent van Gogh's jagged life seemed to me to be incompletely analyzed and obviously unresolved.

The approach throughout this book is to identify the most important questions surrounding van Gogh; wherever possible to start with the artist's own comments and analyses; formulate working hypotheses; look at all the available data as potentially confirming, extending, or challenging; and to present some reasonable conclusions for discussion. In several instances I have tried to place the particular topic in its nineteenth century perspective, to provide some history, and thus include some background on the life and times of Vincent, his colleagues, and his doctors. This is most important with regard to the large advances in medical knowledge that have taken place in the twentieth century. I have also endeavored to summarize previous hypotheses for the sake of completeness and also with the hope of dispelling some myths surrounding this great artist.

Paintings and drawings, especially those facing the opening pages of chapters, have been selected in an attempt to reflect the theme of the narrative. Likewise, the chosen epigraph in each case appealed to me as a variously insightful, poignant, humorous, or poetic example from the artist's correspondence, which fitted the particular chapter.

The line drawings, maps, and graphs were generated on my personal computer. In many of the van Gogh data presentations I have exploited the techniques used in the hard sciences for the illustration of trends, proportions, and temporal displays generally. Also, I trust that the clustering of van Gogh letter references within chapter 3, according to medical symptoms and complaints, will be of value to those readers who are amenable to an organized and more quantitative approach.

The chemical and medical themes that run through the book are vital to my thesis. The goal has been to give as much relevant information as possible, and yet allow different readers to pursue some of these subjects to varying degrees, depending upon background. Accordingly, I have attempted to write with graded and progressive development of each topic for the broad audience interested in Vincent van Gogh.

Wilfred Niels Arnold Ph.D.,
Westwood Hills, Kansas,
February, 1992.

Chapter 1
Introduction

Figure 1.1. Vincent van Gogh: *Garden in Winter*. 1884, Nuenen, ink and paint on wove paper, 51.5 × 38 cm (20.3 × 15″), Szépmüvészeti Museum, Budapest

*The more my health comes back to normal ...
the more foolish it seems to me ... to be doing
this painting which costs us so much and brings
in nothing ... the trouble is that at my age it is
damnably difficult to begin anything else.*
Vincent to Theo, letter 611, from St.
Rémy, Autumn 1889.[1]

The prominent position of Vincent van Gogh in the art world is now firmly established. In 1990, the centenary of his death, he set another world record for sale of an art work by auction. More important, his paintings and drawings in museums, and reproductions in books, wall-hangings, calendars, and even postage stamps continue to attract, intrigue, and entertain a wide audience. Polls find van Gogh near the top of recognized artists irrespective of the aesthetic background of the survey. The present degree of popularity is immense; during his life it was miniscule.

Vincent[*] did receive encouragement by a significant cadre of colleagues during his ten years as a full-time artist, but general acceptance of his artistic contributions came after death, and even then only slowly. Formal recognition during his jagged life was restricted to exchanges of paintings with other artists, gifts of his work to doctors and friends (received with mixed appreciation), some acceptance of canvases toward the artist's financial obligations, small series of commissions,[**] a drawing sold in The Hague, a few items sold in Paris, a self-portrait sold to a London dealer in 1888, and one sale from a Brussels exhibition of 1890. Posthumous praise of his creations roused attention but surely it has been interest in extraordinary aspects of the person that has made Vincent van Gogh a household name. His short life span (1853–1890), range of trials and tribulations, false starts in various vocations, luckless love affairs, intensity and productivity in his last two years, and episodes of debilitating illness leading to self-mutilation and finally suicide, add up to a story that elicits gnawing sympathy and haunting curiosity even in the most casual observer.

---

[*]   The artist preferred to use his first name alone for professional purposes (letter 471), after the precedents of Michelangelo (Buonarroti), Raphael (Santi), Rembrandt (van Rijn), and others. The paintings he signed (only a fraction of the total) were simply inscribed *Vincent*. I have elected to use Vincent van Gogh, van Gogh, and Vincent interchangeably.

[**] In 1878, Vincent made four biblical maps for his father for 40 francs. Uncle Cor commissioned twelve views of The Hague in 1882 for 30 guilders, and then ordered a further series of six, for which Vincent received 20 guilders. Antoon Hermans of Eindhoven commissioned six large designs for interior decoration in 1884.

It is not unreasonable to suppose that all artists of our century have been influenced to some extent by van Gogh. According to Picasso, "[Vincent] managed to break through and make a broad path to all the possibilities of the future."[2] Vincent also became a model, really an icon, for creative artists. The naive view suggests that van Gogh was completely self-taught, lacked proper funding, and was victim to a miserable life style. This image seems to sustain, by sympathetic comparison, the *struggling* artist who hopes and prays for eventual recognition.

In fact, van Gogh did receive some formal training[*] and was supported for ten years, both practically and emotionally, by his brother Theo. His monthly stipend was not princely, but superior to that of a qualified, primary school teacher of the day.[2] While Vincent was sometimes receiving as much as 150–200 francs per month, the postal worker Roulin had to support a family of five on 135 francs (letter 572). Vincent's deprivation was partly due to his illness, but was also in large part self-inflicted; for example, he elected to spend money on the best of artist's materials and frequent changes in venue rather than on food and clothing. Such expenditures can be rationalized as part of an admirable commitment to a professional career but they do reflect poor judgment, especially in light of the fact that malnutrition turned out to be an important exacerbation factor for his underlying illness. A more sophisticated view, one that leads to a more realistic model for future artists, is that van Gogh presented with many of the attributes of creative people generally, namely commitment, focus, hard work, and a willingness to take chances. There can be no doubt about his work ethic as a practitioner of art.

In 1880 Vincent wrote,"I work regularly on the *Cours de Dessin Bargue*,[3] and intend to finish it before I undertake anything else, for each day it makes my hand as well as my mind more supple and strong; I cannot be grateful enough to Mr. Tersteeg[**] for having lent it to me so generously" (letter 136). Charles Bargue was an accomplished but minor French painter who turned

---

[*] Vincent received drawing instruction in high school at Tilburg (1866–1868), a few two-hour lessons in perspective from an unidentified professional artist in Brussels (Winter,1881), structured training in The Hague (January–March, 1882) from his cousin Anton Mauve, a short academic exposure in the School of Fine Arts, Antwerp (January–March, 1886), and critiques at Fernand Cormon's studio, Paris (May–July, 1886).

[**] Hermanus G. Tersteeg managed the Goupil Gallery in The Hague, which had been built by Vincent's Uncle Cent. Vincent's early employment was with the Goupil offices in The Hague, London, and Paris. It is worth mentioning that Adolphe Goupil, the founder of Goupil & Cie, was also instrumental in bringing Charles Bargue and Léon Gérôme (Goupil's son-in-law) into collaboration, and publishing their text.

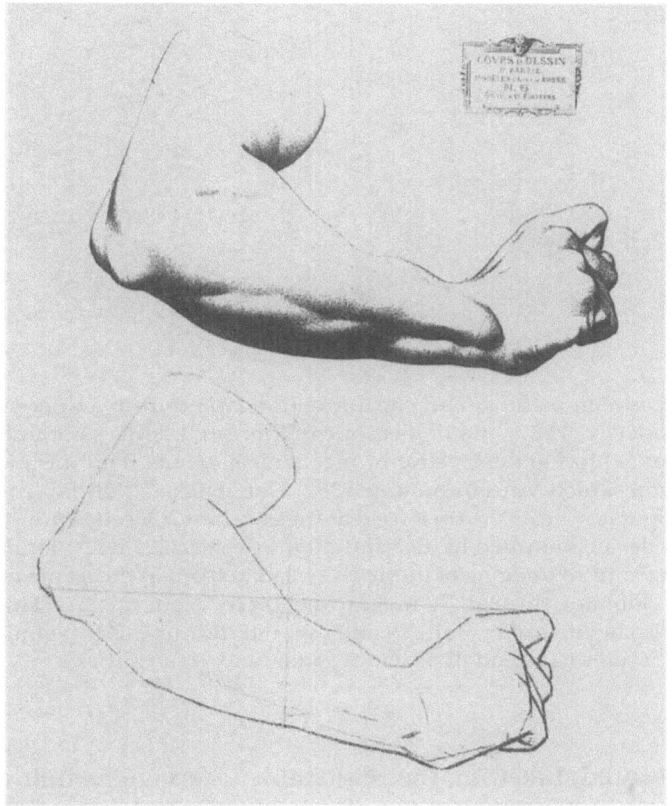

Figure 1.2. Drawing of the arm, illustration after Charles Bargue, *Cours de Dessin*, 1868–1870, courtesy Dr. L. Loftus.

to lithography after about 1867 and, in collaboration with J.-L. Gérôme, published his exercises in drawing, a three volume, large-format series of isolated studies of arms, ears, and other parts of the body, (Figure 1.2), which gradually led to the full torso and then ensembles.

Vincent went through the complete course three times. And then, ten years later, at the height of his career, he requested Theo to "send me Bargue's *Exercises au fusain* ... I need it urgently" (letter 636). In essence, he wanted to refresh his technique. Such diligent application to the basics contributed to and sustained his overall ability. The acquired skill for rapidly pulling off drawings and paintings undoubtedly contributed to the feeling of spontaneity that he was able to engender in so many of his masterpieces. The initial support from H. G. Tersteeg in particular, and the Goupil management in general, was reasonable and laudable. The image of Vincent being completely shunned by the commercial art establishment has

Figure 1.3. Perspective apparatus. The illustration is adapted from a woodcut, "Draftsman drawing a nude,"[c. 1527] and depicts one of four such devices recommended by Albrecht Dürer for achieving perspective by mechanical means. This was probably the model for the frame which Vincent built in 1882. Note Dürer's sighting rod, which is fixed before the operator's eye; the tip is used as the constant reference point. Vincent's sketch of his own device mounted in the sand dunes (letter 222) does not incorporate a sighting rod. Inadvertent dodging of the artist's head in front of the frame may explain his remark that "looking through it makes one dizzy" (letter 223). This quirk in application may also be the origin of the fleeting yet slightly unsettling feeling about the perspective which seems to attend first view of some van Gogh canvases.

been misrepresented. Likewise, the reasonable interactions that van Gogh maintained to the end with a significant number of professionals tend to be overshadowed by the popular image of strained and even violent relationships.

Vincent also mentioned "reading a book on anatomy, and another on perspective, which Mr. Tersteeg also sent me" (letter 136). In this instance, the neophyte artist was delving into the writings and illustrations (Figure 1.3) of Albrecht Dürer,[4] which were influential. Later he wrote that he was "making an instrument for studying proportion and perspective," (letter 205) and "I ordered a new and I hope better perspective frame, which can be fixed in uneven ground in the dunes by two poles," (letter 222). After further embellishment from the blacksmith it was "a fine piece of workmanship ... *It cost me quite a lot*" (my italics, letter 223). Van Gogh appreciated good advice or instruction and willingly acknowledged this help in his letters; there are numerous references to predecessor painters, contemporary artists, past and present philosophers and writers – all of his role models in the informal college of creativity.

Van Gogh's prominent place in all modern analyses of creativity has aroused considerable interest in his medical problems, as well as conjecture about their possible influence on his work as well as his life. The illnesses of creative people justifiably intrigue us. For centuries we have wondered whether sickness creates a new level of awareness in the suitably endowed artist. Likewise, questions continue to surface about the influence of substance abuse on the creative process, particularly whether alcohol and other drug excesses change one's view of the world. The specific case of Vincent van Gogh certainly entertains all of these possibilities. His success compels analysis. The depth of his own written record makes it possible.

The preservation, compilation, and translation of correspondence among Vincent, his brother Theo, other family members, colleagues, and friends was largely due to Theo's widow, Johanna van Gogh-Bonger. The English version[1] consists of volumes I through III with 559, 625, and 625 pages, respectively, and weighs about 10 pounds. The first grouping (about 82% of the total) consists of over 650 letters by Vincent, mostly written to his brother Theo, but interspersed chronologically with occasional missives to his mother. There are three smaller sections of letters by Vincent toward the end of the third volume: 58 to Anthon van Rappard, painter and friend to Vincent from 1879; 23 to Vincent's sister Wilhelmina; and 22 to another painter and long-term friend Émile Bernard. Unfortunately, only 39 letters from Theo to Vincent appear in the *Complete Letters*. Vincent sometimes wrote to his brother more than once a day, and also out of turn, but it is clear that less than 6% of Theo's steady correspondence was preserved. It is generally agreed that this low rate was due to Vincent's neglect rather than posthumous losses.

The collected letters include lengthy discourses on subjects ranging from art history through color theory to personal philosophy. Vincent's appreciation of other artists, his assessments of the literature of the day, an altruistic concern for the future of the general artistic enterprise – all make interesting reading. Some of the items are lengthy: for example, letter 133 is 6.5 printed pages or close to 5,000 words in the English translation. Imagine Theo's smile, on that July day in 1880, when he took receipt of the veritable sheaf of hand-written pages!

Vincent started writing to Theo with some regularity in August 1872, and their exchange continued until his death in July 1890. The time course of Vincent's correspondence is shown in Figure 1.4. The data points are cumulative scores on the number of letters; the intervening lines are labelled with the locales from which each increment originated. The slopes of these lines give some indication of letter writing intensity, which was notably high during Vincent's second period in The Hague and in Arles. The sparsity of letters from Paris, a plateau on the curve, reflects Vincent's cohabitation with Theo.

These letters constitute a cultural contribution in their own right. They complement the art work but also reveal the evolving philosophy of a

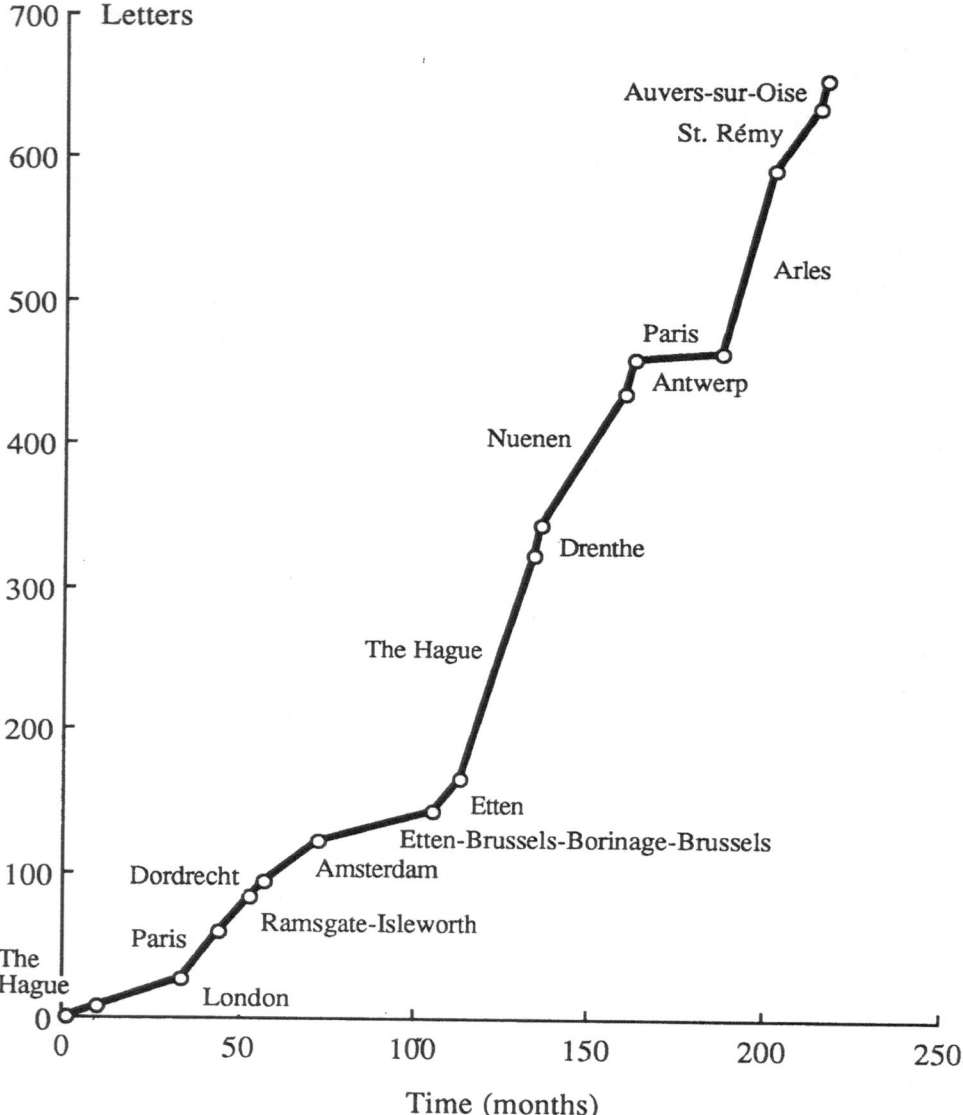

Figure 1.4. The Letters. From August 1872 till his death in July of 1890 Vincent van Gogh wrote a great number of letters to his brother Theo, other family members and friends. More than 650 of these have survived. The time course of this correspondence is depicted. Each point represents a cumulative letter score, for the elapsed time in months after August 1872. Each line joining successive points is labelled with the city, town or district from which the increment of letters originated. The slope of the line provides a measure of the letter writing intensity at each location.

creative artist. The letters were highly regarded by Vincent's brother Theo, who expressed interest in publishing some of them as early as 1890. We suppose that Theo, and the other recipients of Vincent's letters, were variously and intermittently amused by his recall of daily events, elated by his episodes of manic enthusiasm, depressed by his stints of suffering, saturated with his whining, or frustrated by their own inabilities to ameliorate his condition. And yet we readers (once removed) are now mostly buoyed by van Gogh's stream-of-consciousness style and impressed by the depth of self-analysis. The letters also constitute the best source of unembellished descriptors of his medical problems, albeit in lay terms.

The elements that contribute to the mystique about Vincent, and the titillation that attends recall of some of his more bizarre acts, are obvious in most of the popular press, cinema, and television exercises about the artist's life. The highlights have been elevated to more or less common knowledge but too often they engender a simplistic view which lacks accuracy, compounds the myth, and constructs an inappropriate icon. But, they also arouse frequent and valid questions about the nature of van Gogh's underlying illness: why he cut off part of his left ear, why he committed suicide, and whether the episodes of sickness were a prerequisite for, or contributed to, the style and content of his canvases.

A few lonely commentators have denied the necessity to explore these questions. However, the vehemence with which they resist scientific enquiry and communication suggests an unwholesome desire to maintain the mystique in order to protect commercial or special interests in the art. They do an injustice by assuming that the "consumer" of art needs "protection." On the contrary, I believe that explanations of the roles played by his underlying illness and the environment will enhance rather than diminish genuine interest in van Gogh's creations. The results of investigations of medical aspects of the artist and his family may seem to be iconoclastic in the short run but will provide a more realistic and happier model for the long term.

In the chapters that follow I endeavor to analyze the focal questions introduced above, and more, and I attempt to provide working hypotheses with chemical and medical supports. Vincent van Gogh was not insane. He suffered from an inherited, debilitating disease, which was unrecognized in his day. His life style provoked symptoms, exacerbated his condition, precipitated acute attacks, and shortened his life. His illness affected his work. The creative genius is recognized in Vincent's art; the extent of the accomplishment is testament to courage and hard work in spite of adversity.

References and notes

1. The epigraph comes from letter 611, which begins on page 224, volume III, of *The Complete Letters of Vincent van Gogh*. 2nd ed., 1978. Boston: New York Graphic Society. Vincent van Gogh rarely dated his letters. Although they were assembled in chronological order by Johanna van Gogh-Bonger, exact dates are not always available. Henceforth, for the sake of brevity, all references to *The Complete Letters* will be noted in the text, parenthetically, by letter number.
2. Tralbaut ME. 1981. *Vincent van Gogh*. New York: The Alpine Fine Arts Collection Ltd.
3. Bargue C. 1868–70. *Cours de Dessin*. (Course of drawing.) Paris: Goupil & Co.
4. Dürer A. 1538. *Unterweysung der Messung, mit dem Zirckel und Richtscheyt*. (Teaching of measurement with compass and straight edge.) Revised, posthumous edition. Nuremberg: Hieronymous Formschneyder (Andreä). See also:
   Scherer V. 1904. *Dürer des Meisters Gemälde Kupferstiche und Holzschnitte*. (Dürer, the masterpieces of painting, engraving and woodcut.) Stuttgart: Deutsche Verlags-Anstalt.
   Panofsky E. 1955. *The Life and Art of Albrecht Dürer*. Princeton: Princeton University Press.

Chapter 2
Vita

Figure 2.1. Vincent van Gogh: *Entrance to the Public Gardens in Arles*, 1888, Arles, oil on canvas, 72.5 × 91 cm (28.5 × 35.8″), The Phillips Collection, Washington, D.C.

*Either inside or outside the family, they will
always judge me or talk about me from different
points of view, and you will always hear the
most divergent opinions about me. And I blame
no one for it, because relatively few people know
why an artist acts as he does.*
Vincent to Theo, letter 142, from
Brussels, April 2, 1881.

In the little town of Zundert, in the southwest of Holland towards the
Belgium border, Vincent Willem van Gogh was born on March 30, 1853,
and named for his two grandfathers (Figure 2.2). He was the first[*] surviving
child of six to Theodorus van Gogh, an itinerant Calvinist pastor from
Benschop, and Anna van Gogh, *nee* Carbentus from The Hague.[1,2] Vincent
died in Auvers-sur-Oise (France) on July 29, 1890. This was just five years
and three months after the passing of his father, who was disappointed with
his eldest son's failure to bring any of four previous vocations to fruition
and was unaware of his potential as an artist.

The Reverend Theodorus van Gogh was sixth among twelve children
(one died in infancy) born to Vincent Johanneszn van Gogh, from The
Hague, and Elisabeth van Gogh, *nee* Vrijdag, from Rotterdam.[2] His theological studies in Utrecht were interrupted by a serious but undocumented
illness. He was again seriously incapacitated in 1843, at age 21, after a long
walk. At least one commentator[3] has judged the pastor to have been in less
than good health most of his life. Theodorus van Gogh died at age 63, from
a stroke.

Vincent's mother, Anna, survived all three of her sons and lived long
enough to witness the beginning of public interest in van Gogh's art. She
was one of eight children born to Willem Carbentus and Anna Carbentus,
nee van der Gaag. Her father was a bookbinder and bookseller in The Hague.
Anna's sister Cornelia also married into the van Gogh family, to Uncle
Vincent, who started a modest shop in The Hague for artist's supplies and
built it into a large, financially successful art gallery. Uncle Cent, as he was
called by Vincent and his siblings, amalgamated with Goupil & Cie [Co.] of
Paris in 1858, but suffered from poor health to the extent that he was
regarded as too weak to go to college,[4] and took early retirement at age 53.
The double familial attachment to Anna and Theodorus van Gogh, which
was nurtured by residential proximity (Uncle Cent and Aunt Cornelia
settled in Prinsenhage), was largely responsible for Vincent's obtaining a
position at Goupil's, and the older man's stature in the company helped

---

[*]   Another child, also named Vincent Willem, was stillborn in 1852.

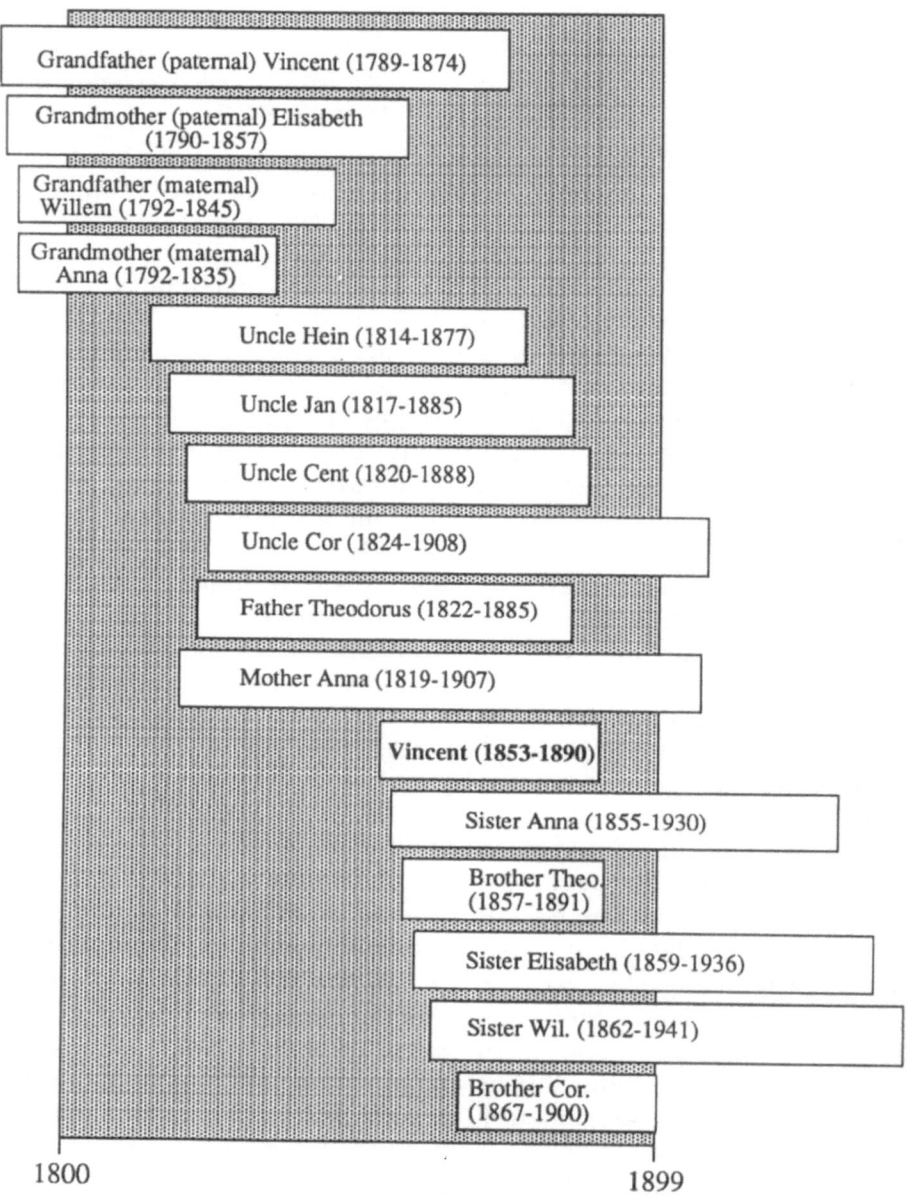

Figure 2.2. Three generations of van Goghs. The horizontal axis is scaled in calendar years; the nineteenth century is stippled. The box for each family member is bounded on the left by his or her birth year and on the right by the year of death.

Vincent survive there for close to seven years. Later, this Uncle was also instrumental in helping him find a job in a bookshop. As we shall see later, he was never impressed with the prospect of his nephew's joining the clergy.

Uncle Cent was childless, had amassed a fine art collection, and was reasonably well to do. Johanna van Gogh-Bonger suggested[1] that he was in the best position to have sponsored Vincent's belated art career. It was not to be; after 1877 there was a marked loss of confidence in the unruly nephew. Yet even in Cent's final year Vincent could reminisce about his favorite uncle, remark about positive influences, and express dismay at knowing someone who had been so active now "reduced to ... suspicious helplessness and continual suffering" (letter 512). The Goupil & Co. which Uncle Cent had introduced in Vincent's youth was "the finest, the best, the biggest in the world" (letter 332).

Two other uncles on his father's side were prominent in the art world. Uncle Hein [Hendrik Vincent van Gogh] had shops in Rotterdam then Brussels, the latter being quite successful and eventually taken over by Goupil. Uncle Hein retired at age 58 for undocumented health reasons, and died five years later. Uncle Cor or "C.M.," Cornelis Marinus van Gogh, founded his own art company in Amsterdam in 1862; this house was also influential in Holland.[3,4] The artistic environment created by the extended family, Vincent's exposure to art within Goupil & Cie as well as his visits to some of the best museums of Europe, and his rubbing shoulders with artists and connoisseurs – all played an inestimable role in the future development of Vincent van Gogh as a creative artist. These exposures complemented his meagre years of primary and secondary schooling.

Vincent attended a village school in Zundert, and then in 1864 transferred to a private boarding school in Zevenbergen, where he stayed for two years. He apparently made sufficient progress to be admitted with good standing to the new, well-endowed, State Secondary School in Tilburg, where he received superior instruction in Dutch, German, French, English, arithmetic, history, geography, botany, zoology, drawing, calligraphy, and gymnastics.[5] This period lasted a little less than two years. Formal schooling, perhaps a solid high school education by today's standards, was over at age 15. His reasons for not finishing the last term, returning to Zundert, and not taking up employment for another year have not surfaced.[5]

Art dealer

On July 30, 1869, Vincent became a junior clerk with the branch of Goupil & Co. in The Hague. This commercial gallery had been founded by Uncle Cent but was now managed by Hermanus G. Tersteeg. Vincent's performance was judged more than adequate and he received at least one raise in salary during this period of close to four years. Much later he declared that "the first two were rather unpleasant, but the last one was much happier" (letter 266). Armed with a good recommendation from H.G. Tersteeg, Vincent transferred in mid-May of 1873 to the London office of Goupil.

Much has been written about Vincent's evolving attitude toward the management as well as their clientele. He received another raise in London but the once obliging clerk changed into a judgmental and verbal connoisseur, good for personal growth but not always good for business. His preoccupation with activities outside those of his employ was a recurring theme in the years before he dedicated himself to painting; for instance, reading English literature while working in the London gallery, perusal of the Bible during the last sojourn in the Paris gallery, being more inclined to talk about either art or religion than books in the Dordrecht bookstore, absorbing art instead of Latin while preparing for the ministry in Amsterdam, and writing about paintings rather than religion during his studies in Brussels. The sympathetic observer can rationalize Vincent's interests but he was a trial for employers and teachers.

During the London period Vincent suffered his first experience of love. The object of his somewhat Platonic attention was Eugénie Loyer, a daughter to the landlady. She was hardly aware of the relationship, which was short-lived, one-sided, and unproductive. It was one of at least three[*] unhappy, clumsy, and inept affairs spread over his lifetime. It contributed to Vincent's discomfort with London, and was discussed in depth when he visited his family, now in Helvoirt,[**] towards the end of June 1874. His oldest sister, Anna, accompanied him back to London, perhaps in part as a stabilizer but mostly to find her own life. Later that same year, Vincent visited the head office for three months, supposedly for reorientation. Much to his displeasure he was permanently assigned to Paris in May 1875.

After several discussions involving his father, Uncle Cent, and Mr. Boussod (the company was now called Boussod, Valadon & Co., successors to Goupil & Co.) it was mutually agreed that Vincent should resign effective April 1, 1876. The immediate cause of the employers' chagrin was Vincent's going home (now Etten) at Christmastime, when gallery sales were usually brisk. But the sources of dissatisfaction were cumulative and mutual. Vin-

---

[*]   The others involved his widowed cousin Kee Vos-Stricker from Amsterdam, and a neighbor in Nuenen, Margot Begemann.

[**]   Pastor van Gogh was moved by the church leadership from one small town to the next within the Noord Brabant. He had small congregations, and was noted for pastoral rather than preaching skills. His work was always judged satisfactory, but he was never called to the big cities of the north with Protestant majorities.

cent returned to Etten in March, 1876. He had already found his own way from a newspaper advertisement: he was to be a teacher in a boys' school located at Ramsgate, on the east coast of England.

School teacher

It was a nonsalaried position; the headmaster, Mr. Stokes, provided only room and board. A few months later the establishment was moved to Isleworth, to be closer to London. The job had no future and Vincent started to think about becoming a clergyman, or at least an evangelist, but was surprised to learn that he was too young. He did receive a small stipend from a Reverend Jones for reading Bible stories to young children at nearby Turnham Green. This opportunity proved fulfilling, and grew into an assistantship in the Methodist church in Richmond. Vincent offered a couple of sermons which by his own admission were better read than heard. He maintained friendly relations with Rev. Jones but was homesick and, after returning to Etten for Christmas, tendered his resignation. There was nothing ignominious about the separation and it is worth remembering that Jones took an active interest in Vincent's later evangelical aspirations.

Bookseller

After a short break (the unstippled sectors of the occupations chart, Figure 2.3, denote periods of inactivity) Vincent assumed a position in January 1877, with Blussé and Van Braam, booksellers of Dordrecht, Holland. He was a clerk for about 13 weeks. His duties were keeping records and giving advice on artistic prints rather than actually waiting on customers. The hours were long and the work was boring. He disappeared into a back room and developed the unsanctioned hobby of translating passages of his Bible into French, English, and German. While in Dordrecht he carried out a mission of kindness to his birthplace in Zundert to visit a sick farmer, spent the night in the churchyard, experienced a warm feeling about the pastoral work of his father, and wished to emulate him. The notion of theological study entered his head and just had to be addressed. After a short visit to Etten he was ready to undertake preliminary private lessons in Greek and Latin; he would live in Amsterdam after May 1877, with his Uncle Jan, and seek advice from Uncles Cor and Johannes (brother-in-law to Vincent's mother).

28

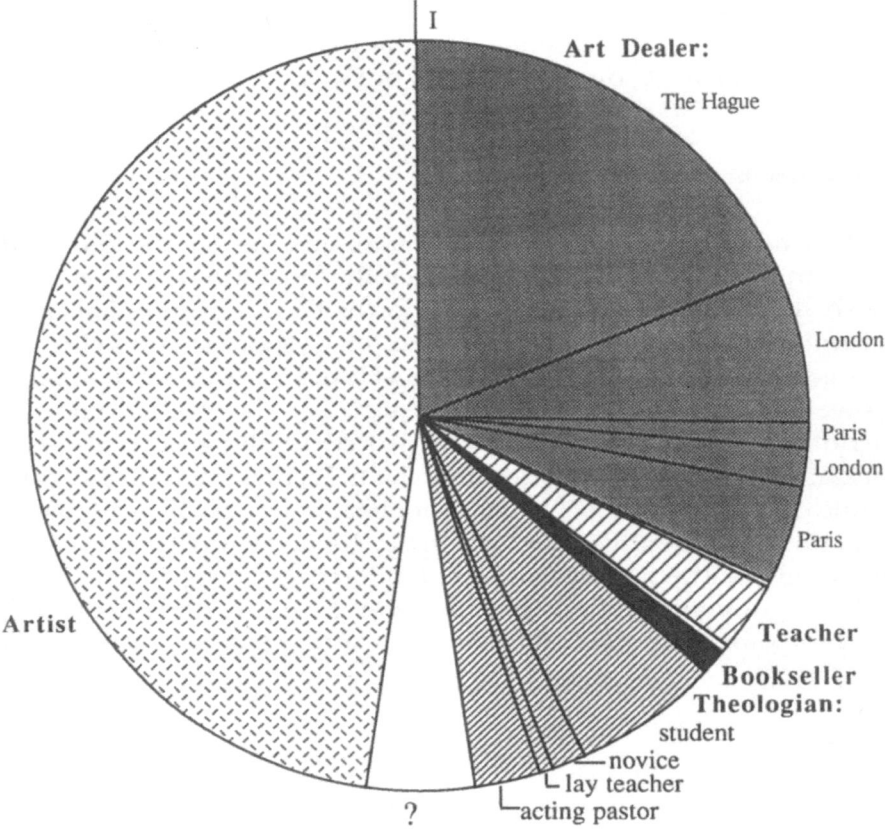

Figure 2.3. Occupations. The chart indicates the order [clockwise from the top], and the relative times, which Vincent spent in five major vocations after high school; namely, art dealer, teacher, bookseller, theologian, and artist. The initial period of employment with Goupil & Co. is sub-divided according to branch location and chronological order. Sub-categories within theology are also indicated. The total period was a little over 21 years of which slightly less than 50% was as an artist.

Theologian

The Reverend Johannes Stricker was a noted preacher in Amsterdam and he gave Vincent some private lessons on the doctrine of the church. He also recommended Dr. Mendes da Costa, a scholar of classical languages, who was pressed into service to assist in preparation for the state university entrance examination. According to Vincent's sister Elisabeth, "During the time the young man [Vincent, age 24] was preparing for his examinations he had a mental breakdown ... he not only spent half the night working but wrote incessantly [until] ... the writing was no longer legible ... and became mere pen-strokes." Unfortunately this reference cannot be taken at face value; Elisabeth du Quesne-van Gogh did not prove very reliable in her recollections,[6] at least with respect to dates and places.

It would have taken Vincent seven years to become fully qualified in theology; he lasted one year in preparative studies; he did not take the test. Given Vincent's command of English, French, and German it is surprising that da Costa found infertile ground for Greek and Latin. A more likely explanation is that Vincent was displeased with the requirements, impatient to enter pastoral activity, and that he revolted against the system. Later he declared that he was "very skeptical about the plan of studying ... might have managed to master that miserable little bit of Latin ... [but it was] too much for me ... [because] the theological faculty [at the university] ... [was] an inexpressible mess, a breeding place of Pharisaism" (letter 326). Unfortunately, Vincent's next step was also poorly conceived and badly executed.

In mid-July 1878, Vincent, his father, and the Rev. Jones of Isleworth, approached the School of Evangelization in Brussels for entry into "a three-year course ... [requiring] less knowledge of ancient languages ... [and valuing] more highly [my] fitness for practical work" (letter 123). The trio seemed to turn a blind eye to the time frame.

A three months' probationary period as a novice in the Laeken school, near Brussels, was terminated by Vincent undertaking Bible-teaching further to the South, in the Borinage district. In Pâturages he also volunteered to visit with the sick, and in January 1879 was named acting pastor of Wasmes, near Mons. The Rev. Theodorus van Gogh visited the Borinage in March (letter 128) and was utterly dismayed with his son's lifestyle, which included renouncing the comforts of even modest living quarters, and donating to the poor miners "the clothes off his back." The father seems to have reached his limit concerning the downward spiral of Vincent's standing within the theological ranks. The discomfort of the son was compounded by nonrenewal of his contract after June 1879. He stayed in the Borinage region and thereafter his address was either Wasmes or Cuesmes, both near Mons. Disillusioned with formal religion, and torn between evangelism and art, Vincent van Gogh entered a period that is best described as wandering. The time line (Figure 2.4) is a graphic summary of frequent changes in residence during Vincent's life.

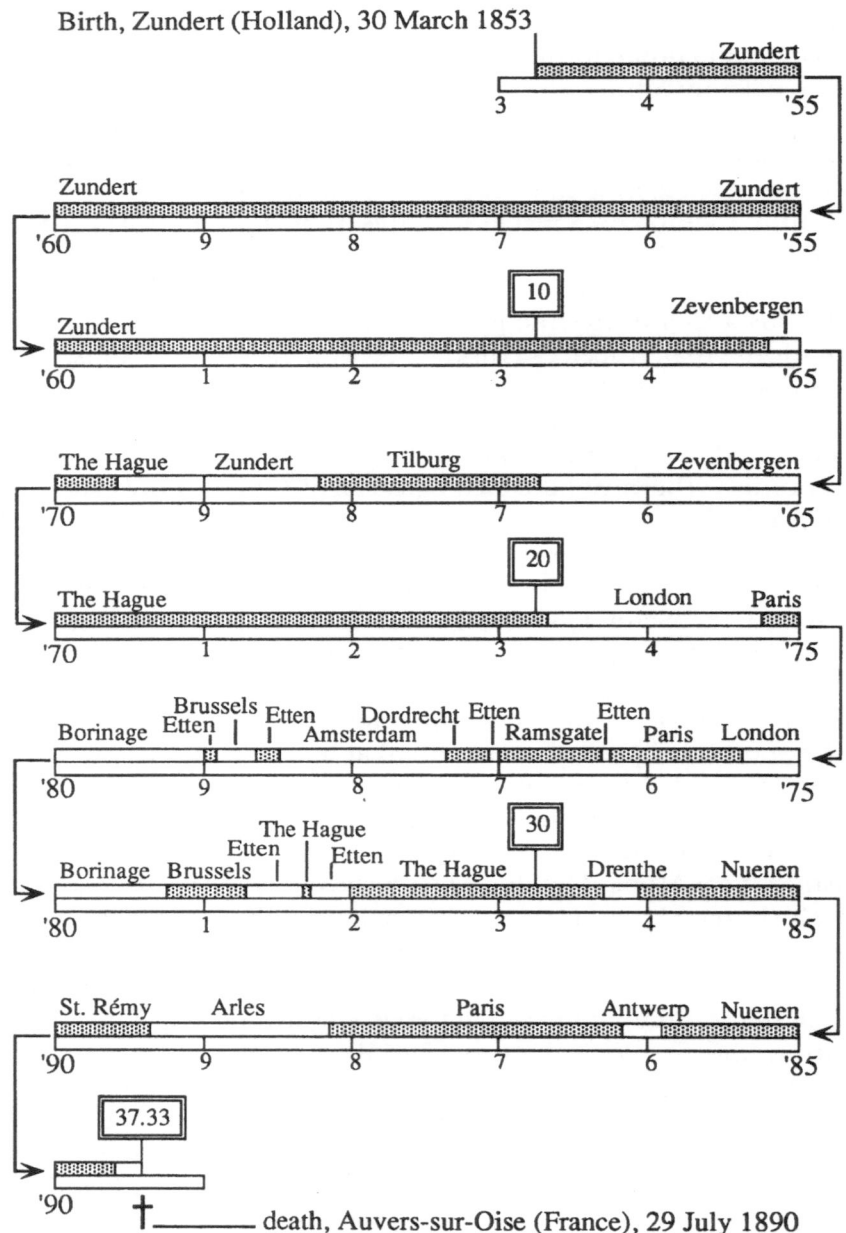

Figure 2.4. Time line. Vincent van Gogh had 30 changes in residence during his life of 37 years and 4 months. The locations are indicated by alternating stippled and clear areas along the upper track; the calendar years are mapped on the lower track. There are boxed markers for van Gogh's birthdays at 10, 20, and 30, and his decease at 37.33 years of age.

Wanderer

The middle of August, 1879, found him back in Etten. His parents thought he looked remarkably well except for his mode of dress, which they attempted to correct locally. Vincent spent all of his time reading Charles Dickens, Harriet Beecher Stowe, Victor Hugo, and Jules Michelet. On one occasion father and son walked together to Prinsenhage to see Uncle Cent's paintings but, under steadily increasing animosity on the home front, Vincent returned to the Borinage and lived on irregular gifts from his father and brother Theo.

Sometime in March 1880 Vincent walked all the way to Courrières, near Lille, France, to seek an audience with Jules Breton, an admired painter, and to look for work: he was unsuccessful in both missions. On the return he was broke, slept anywhere he could find shelter, and exchanged drawings for food.

In the Spring there was some discussion among the family about setting Vincent up in a store for artist's supplies, with the view to also selling art of his own taste, something along the lines of Uncle Cent's initial venture. Neither the capital nor the commitment was forthcoming. Vincent went back again to the Borinage and wondered out loud, "How can I be useful, of what service can I be? *There is something inside of me, what can it be?*" (letter 133, July 1880; my italics). In the same letter van Gogh made at least three additional and important points; first, the comparison between literature and visual art, "the love of books is as sacred as the love of Rembrandt – I even think the two complement each other. ... how beautiful Shakespeare is! ... His language and style can indeed be compared to an artist's brush, ... one must learn to read, just as one must learn to see and learn to live." Second, about religion he wrote "with evangelists it is the same as with artists. There is an old academic school, often detestable, (etc.) ... everything which is really good and beautiful – of inner, moral, spiritual, and sublime beauty in men and their works – comes from God, ... the best way to know God is to love many things. Love a friend, a wife, something – whatever you like." And third, he declared that "A very necessary study is that of medicine; there is scarcely anybody who does not try to know a little of it, who does not try to understand what it is about, and you see I do not yet know one word about it."

In this mammoth communication (letter 133) Vincent admitted to writing "somewhat at random whatever comes to my pen," and I believe that the thoughts expressed here set the stage for the future. They served to rationalize for himself, as well as Theo, the substitution of artistic creativity for evangelical work, and he proposed that the two activities were equally acceptable to a supreme being. He anticipated his affair in The Hague as an attempt at loving "something – whatever you like," or perhaps we might now say, whatever you can. It excused his wandering time, which his family called "idle" as educational, rather well spent, and necessary because he asked, "How [else] can I learn more and study certain subjects profoundly? ... that is what preoccupies me constantly."

Artist

Although others have argued for a somewhat earlier start, I agree with Johanna van Gogh-Bonger's assessment that letter 134 is a clear indication of a watershed, Vincent's declaration to be an artist. It was written from Cuesmes, and has been dated August 20, 1880. Vincent wrote about busily sketching large drawings based upon depictions of working country folk by Jean Francois Millet. He also requested more prints from Theo, and mentioned writing to Mr. Tersteeg for Bargue's *Cours de Dessin*. Eighteen days later he announced completion of ten sheets after Millet and 60 sheets of Bargue's exercises (letter 135). He was off and running as an artist.

A sudden move to Brussels, to be near art and artists and to have more room (letter 137), occurred in October 1880. Vincent gravitated to the branch of the Goupil & Co. originally owned by Uncle Hein. It was now administered by a Mr. Schmidt who was cordial, in spite of some financial litigation with Uncle Hein (letter 138). Schmidt recommended the École des Beaux-Arts. Vincent also followed Theo's suggestions and contacted Willem Roelofs, a successful Dutch painter living in Brussels, and Anthon van Rappard, a serious but unfulfilled artist who had studied in Academies of Amsterdam and Paris before registering in Brussels. They all urged Vincent to join the Academy and he was pleasantly surprised to find that any student who could procure endorsement from the mayor would be granted free tuition. It has been assumed by most commentators that Vincent applied for this dispensation but was unsuccessful; however, there is ample evidence from letters that he was more interested in books and self-study. He continued with the exercises of Bargue and also took some lessons in perspective from a professional artist who has not been identified. The six months in Brussels were artistically bright, but living conditions were dim.

By January 1881 Vincent contemplated a move to The Hague, while wondering about the potential for receiving more help from Tersteeg, who had already sent him anatomy books and the Bargue manual, and from his cousin Anton Mauve (letter 139). In the next letter he mentioned drawing *for the third time* all the exercises of Bargue. Meanwhile, brother Theo was busy, and prospering, within the Paris branch of Goupil and Vincent returned to an earlier idea of his coming to France. There is even a suggestion (letter 140) of his hoping to find a salary of 100 francs a month; could this be interpreted as a plaintive cry for reinstatement with Goupil?

His father visited once and they apparently discussed the reluctance of the uncles and Mr. Tersteeg to assist Vincent in the manner he sought. The financial contributions by Theo, which had been sent through his father, were now openly revealed to Vincent for the first time. The father's chagrin was increased by the perception that Roelofs, Schmidt, and even van Rappard found it difficult to reconcile Vincent's wretched life style with the relative success and comfort of his relatives. In short, Vincent had become an embarrassment to the family!

Etten

The immediate reason for Vincent's departure from Brussels in April was the opportunity to see Theo back in Etten. After a day in bed and a long talk with Dr. Van Gent about his "insignificant malaise," Vincent decided to stay in Etten to enjoy regular meals and larger quarters than any since Amsterdam. He continued to draw, and was visited twice by van Rappard. In September of 1881 Vincent visited The Hague and was encouraged by Anton Mauve, who suggested that he was ready to start painting (letter 149).

According to most commentators, the event of the year was Vincent's unrequited love for a widowed cousin, Kee Vos, the second daughter of the Rev. J.P. Stricker. Vincent was more than attentive to Kee, and her eight year old son, while they visited the parsonage in Etten that summer. In the fall Theo sent money so that Vincent could visit the lady at home and make a renewed plea. It is of interest in terms of priorities to note that en route to Amsterdam Vincent stopped off for a month in The Hague to receive instruction from Mauve, and that he spent the bulk of the stipend on affairs of art rather than heart. As it turned out, Vincent was refused audience with the girl, banished by the angry father, and relieved by a prostitute. During the abortive visitation Vincent mutilated his hand; much later he described this to Theo, "I put my hand in the flame of the lamp and said [to J.P. Stricker], 'Let me see her [Kee] for as long as I can keep my hand in the flame' ... But I *think* [my italics] they blew out the lamp" (letter 193). The consequences were also felt at home and contributed to his father's decision to kick him out on Christmas day (referenced the following year, letter 166).

The Hague

He went to The Hague, which fulfilled his yearlong wish, and obtained quarters not far from Mauve. He considered the beach town of Scheveningen but was put off by the relative expense. This was one of the few financially reasonable decisions he would make during the next twenty-two months. There were other early indications to suggest that this might be a joyous and productive period.

However, the first bout of serious illness was encountered towards the end of January 1882 (letter 172). Vincent was, "so miserable that I went to bed; I had a headache and was feverish" (letter 172). This episode lasted three days, was accompanied by nervousness, toothache, and "heavy depression" and was attributed to "overexertion" (letter 173). The same letter contains a passing reference to Theo being ill. It is worth noting that Vincent was almost 29 years old, and that Theo had just turned 25. Vincent complained that up until a few years earlier he was strong and never had to stay in bed a day, but now there was always something the matter with him. The next week he recovered sufficiently to resume his art work.

The immediate response from Mauve in particular, but also from several other artists, was social cooperation and a willingness to share techniques. Van Gogh made progress with his drawings. Mr. Tersteeg bought a small one for 10 guilders (letter 176), but Vincent was not much impressed with the item and suspected a disguised act of charity. The stars began to shine when Uncle Cor commissioned twelve small views of the city for 30 guilders.

The letters were now filled with concern for support. Vincent's constant bickering with Theo, mostly on amount and frequency of stipend, bordered on harassment. The younger brother was prepared to send about 100 francs a month for bare essentials but Vincent favored more liberal spending. He had restocked his own wardrobe, engaged models for days on end, purchased garments to embellish them as subjects, and felt obliged even to feed his sitters. As a consequence of poor budgeting Vincent was continually behind in meeting obligations. He developed the wretched habit of borrowing money from Tersteeg against anticipated allowance from Theo, and then seemed surprised at the loss of confidence of both. He strained his friendships with Mauve and Uncle Cor.

The commission from Uncle Cor had been a "ray of hope" (letter 181). He finished a dozen small drawings in a few weeks. Vincent's reaction was curiously suspicious. Would Uncle honor the arrangement and, after all, "I had trouble enough making them, perhaps more than 30 guilders' worth" (letter 183). In fact, his Uncle was so pleased that he ordered six more (letter 184); these took longer to complete. Vincent delivered seven pieces including two large drawings which today are considered among van Gogh's best.

The artist hoped "that C. M. continues to order" (letter 202), but he received 20 guilders and no comment (letter 205). His relationship with Mauve deteriorated. According to Vincent, "Mr. Tersteeg poisoned Mauve's ear by saying, 'Be careful you can't trust him [Vincent] with money' " (letter 191); the interference was onerous for the artist but not completely unfounded.

There was, of course, more to it than that: Vincent had taken up with a pregnant prostitute, Clasina (Sien) Hoornik, who was three years his senior. Her four year old daughter, ten year old sister, and mother, rounded out the new family (letter 192). They served as free models for Vincent but the added expense of support was too much for Theo. Vincent declared that "My feeling for her is less passionate than what I felt for Kee last year, but the only thing I am still capable of ... is the kind of love I have for Sien" (letter 204). In the meantime the artist had a fourteen day bout of illness including insomnia with fever and nervousness (letter 200).

Theo became stressed by the behavior of his older brother although he never came close to forsaking him. On top of all, Vincent made the seemingly fatuous suggestion that Theo should also become an artist; this was not just a passing comment but an idea that recurred in several letters. How in the world would they have supported each other? Theo seems to have been a paragon of patience and understanding.

As her pregnancy progressed Sien had problems, and Vincent dutifully assisted her to a hospital in Leyden for a correction in the position of the fetus (letter 192). He continued to make plans to marry her. Because of her background this would have been no less attractive to the pastor and his family than cohabitation. Theo tried to talk him out of it but met nothing but resistance. Sien and her mother then moved into their own small house with the plan that the younger woman would later return, with her new baby, to Vincent.

In June 1882 Theo was surprised to receive a letter from the fourth-class ward of the city hospital in Brouwersgracht, within The Hague, and to find Vincent complaining of insomnia, low fever, and painful urination during the previous three weeks (letter 206). He was required to pay 10.5 guilders in advance for an anticipated two weeks. He admitted to a mild case of the "clap" (gonorrhea) for which he received injections (irrigations) with alum solution and many tablets of quinine. In the nineteenth century, quinine was prescribed for fevers in general, as well as the malarial type.[7] Washing the urethra with a solution of alum (aluminum potassium sulfate), an astringent, was recommended for gleet (prolonged gonorrhea infection).[7]

Tralbaut[3] has reproduced the patients' register from this hospital; it indicates that Vincent was admitted June 7, 1882; discharged July 1; suffered from gonorrhea; and was treated with injections of zinc sulfate. Zinc sulfate solution was also used as an astringent irrigation.[7] There is also mention of his doctor putting "a catheter into the bladder quickly, without 'ceremony' or fuss" (letter 208). Vincent thought the treatment and food in the hospital were good, welcomed the rest, and enjoyed visits from Sien who came regularly until she was confined in the Hospital in Leyden.

But to the surprise of his doctor things took a turn for the worse as he complained of a dreadful weakness and on June 22, 1882, wondered if there had been a "complication." He was moved to a new ward and also had to pay for another two weeks in advance. What was the nature and possible cause of this paroxysm?

Vincent's father came to see him once, but the visit was hurried and uneventful and "seemed more or less like a dream, as does this whole business of lying here ill" (letter 208). He was back in his studio by July 1, but was required to make regular visits to the hospital to receive a series of painful stretching treatments to the urethra, which were described in detail (letter 209).

The hospitalization had a bizarre corollary. Vincent claimed that attending physicians were willing to attest to his sanity and that he could, if necessary, "get an official statement ... to the effect that [he was] not the person to be sent to Gheel [a mental institution near Antwerp]" (letter 206). This statement is quite startling at first encounter, but information from other letters, taken collectively, indicates that his father had considered having him committed in 1880 (age 27). I believe that most commentators have too narrowly interpreted family concern in 1880. They suppose that it centered around Vincent's life style in the Borinage, and again in 1881 around the abortive affair with his cousin Kee Vos. Hulsker[4] states that letter 158 (November 18, 1881) contains a passage "but I [Vincent] can't believe that a father is right who curses his son and, think of last year [1880], wants to send him to a madhouse," which was omitted from the *Complete Letters*. This throws a different light on Vincent's concern about loss of civil rights and being "put under guardianship" (letter 204; June 1, 1882). The matter takes on added significance with Vincent's plea, "In case of illness, I hope *you* [Theo] would object if they [my parents] tried to make use [take advantage] of my being powerless." Vincent was convinced that his father had been in earnest in 1880 about sending him to the asylum.

In the light of the father's reasonably benevolent treatment, both before and subsequent to this time, I am convinced that the issues concerned more than the questions of manners, decorum, family relations, and social misunderstandings that Vincent discussed in correspondence. He must have exhibited some alarming and psychotic behavior. For this reason I take 1880 (age 27) as the onset of Vincent's serious illness. The "complication" in June 1882 (age 29) to the uncommonly long gonorrheal infection provides more evidence to suggest that problems arose in the latter part of his third decade.

In July 1882 Vincent complained of nervousness and a terrible sensitivity both physical and moral (letter 212) and anticipated another bout in the hospital. He complained of indigestion, poor appetite, fever, and other problems as being chronic during the last two months, but the doctor [who was given Vincent's best watercolor to date] was satisfied that sufficient progress had been made and further treatment was deemed unnecessary (letter 214). He complained occasionally until letter 238. His girlfriend, Sien, returned with her new baby and Vincent had visions of domestic bliss. However, he was beset with visitations by Tersteeg, who was disturbed by the outlook and, to the artist's consternation, felt compelled to involve both his father and Uncle Cent. Vincent's letters to Theo were now filled with moral rationalizations, attempts to discredit Tersteeg, and the promise of a civil marriage to Sien. The union did not take place and there followed a long series of letters in which she is not even mentioned.

Vincent was now convinced that he must work doubly hard to make up for the late start in his artistic career and for the time lost to illness. He made several day visits to Scheveningen by streetcar, to paint seascapes, to experiment with his perspective frame (letter 222), and to put into practice things that Mauve had told him in January (letter 227). He had spent two years devoted to drawings and now felt confident to attempt painting in oils. The first three were described in letter 224 (August 1882). In the same month he read books by Émile Zola including *The Sin of Father Mouret* in which he took great pleasure, and expressed admiration for the "noble figure" of Dr. Pascal Rougon, a theme that would be repeated. (See Chapter 9 for a possible relationship between Zola's book and Vincent's ear-cutting incident; also, see Table 1 for other examples of books read by Vincent.)

Theo's stipend was now euphemistically called "the enclosure." Vincent wondered about buying brushes and artist's colors wholesale but otherwise remained unwilling to slow down on spending and instead suggested that the stipends should come more frequently: he was obviously barely existing from gift to gift. There was another unexpected visit from his father. The family was not hostile to Sien and actually sent her a winter coat (letter 236), but they were ill prepared to accept the situation as permanent. In the meantime Vincent assembled his "hundred studies" with pride but admitted some despondency (letter 240). Theo's letter of November 9, 1882 was lost in the mails. The consternation that followed indicates that Vincent had apparently been receiving banknotes rather than money orders (letter 245); thereafter the communications containing cash were sent by registered mail (letter 247).

During the period in The Hague Vincent's admiration for selected artists, past and present, increased. Millet received special accolades. Van Gogh also worshipped the English masters of the wood-engraved, magazine illustra-

tion and experimented with a relatively new process for transferring drawings into lithography. He conversed with his brother Theo, van Rappard,
and even the latter's father, about publishing "prints for the people," a
project with a built-in nonprofit motive (letter 249) which, to the relief of
the potential sponsors, did not get off the ground. The enterprise was partly
inspired by Vincent's growing collection of prints. For example, by 1883 he
had acquired most of the illustrations from *The Graphic,* even back to its
inception in 1870. He was entranced by one piece from Luke Fildes cap-
tioned "The Empty Chair," which commemorated the passing of Charles
Dickens. This theme was later embraced in painting "Gauguin's chair"
(1888). Vincent even toyed with the idea of seeking employment in England
as an illustrator.

In the beginning of 1883 the relationship with Sien began to unravel.
Vincent blamed her mother. He spent much of the remaining months
taking day visits to Scheveningen to paint and to escape Sien's family. Theo
was busy with his own affair in Paris with a Catholic girl named Marie (see
letter 332), who apparently participated with him in some delightful times
interspersed with episodes in which she was seriously ill and hospitalized.
She was usually identified as "Theo's patient." Vincent played the amateur
psychiatrist (letter 268a). He pontificated, perhaps for his own as well as
Theo's situation, "no matter how good and noble a woman may be ... in the
present society she is in great, immediate danger of being drowned in the
pool of prostitution. What is more natural than protecting a woman, and,
... if circumstances lead to it, ... marrying her. ... [but one marries] the
whole family" (letter 279). "Women and their mothers, ... in my case
nine-tenths of the difficulties which I had with [my] woman originated
directly or indirectly therein" (letter 281).

Vincent feared that Sien would revert to walking the streets, contemplated moving his foursome (minus the mother) to the country, but was
eventually overwhelmed by debts and doubts, and elected to make a break.
Uncle Cor came to the rescue monetarily. A partial reconciliation with
Tersteeg occurred but he was unable to reach Mauve, whom he now judged
to be capricious (letter 299). In September 1883, Vincent left for the
northeast region called The Drenthe, which was occasionally frequented by
Mauve. Certain sections of The Drenthe were also recommended by van
Rappard and Van der Weele, a recent friend and reasonably successful artist,
who had also painted there. Vincent hoped to live more cheaply and to
explore new motifs.

The Drenthe

The train journey was interesting but spoilt by the memory of his departure from Sien, and even more from her little boy with whom Vincent had developed a mutual love. He stayed at a small hotel in Hoogeveen, which was at the crossroads of several canals. Later he moved by barge through the moors to Nieuw Amsterdam, to be in the heather near Germany. He boarded with a farmer but had to return to Hoogeveen periodically in order to receive the mail from Theo and the all important stipend.

There were problems in exchanging French francs in this part of Holland so Theo turned to mail orders. The whole arrangement proved tiresome because Vincent persisted in extracting reassurances about support and in redefining their *joint* venture. It was compounded by a slow mail delivery and a misunderstanding towards the end, slightly less than three months in all, which left Vincent short of cash for several weeks. His father helped out with a small gift.

The Drenthe trip was ill timed because of the oncoming winter, hurriedly planned as usual, and poorly executed. For example, based on rumor he hoped for a chance encounter with Liebermann, a reasonably successful German artist, but not surprisingly never found him. However, it was a period of reassessment and philosophizing, sitting around peat fires, and observing a new way of life in the countryside. Theo was rethinking his relationship in Paris. His older brother kept making strange references to Sien and some of Theo's money was sent to her on one occasion. In part because of Theo's difficulties, both felt and anticipated, with Goupil and Co., the suggestion of Theo becoming an artist resurfaced; this time Vincent's emphasis was on becoming "brother painters" and he remarked on the examples of Ostade, van Eyck, and Breton. An amusing rider suggested that Theo's girlfriend become a painter as well (letter 335). Even some creative financing was suggested but the mathematics were wrong (letter 339), and Theo resisted.

The Drenthe tour should have encouraged sketching but Vincent became preoccupied with his lack of paints. The work rate actually decreased compared with The Hague. But letter writing proceeded at a high rate and included some of his longest communications. One of the most beautiful describes his journey from Nieuw Amsterdam to Zweeloo, by cart with his landlord and back by foot. "That day passed like a dream, all day I was so absorbed in that poignant music that I literally forgot even food and drink – I had taken a piece of brown bread and a cup of coffee in the little inn where I drew the spinning wheel. The day was over, and from dawn to twilight, or rather from one night till the other, I had lost myself in that symphony" (letter 340).

Vincent remained convinced of the attractiveness of the Drenthe region but his hope of returning some day was never fulfilled. He made the six hour walk to Hoogeveen, bought a train ticket with money borrowed from his father, and joined his parents (now in Nuenen) the first week of December 1883. He planned a visit of only a few months but stayed almost two years. There were several reasons, not least of which his mother fractured a thighbone, in mid January 1884. Her injury, sustained in a fall while alighting from a train, concerned the whole family for several months. Furthermore, the attention given her by Vincent helped to divert the animosity that had been building between son and father.

Nuenen

Vincent felt that his father had not made sufficient accommodation about the ouster of Christmas 1881, or to the events of the previous year. In particular, the artist's relationship with Sien was not properly resolved and Vincent even restated his intention of marrying Sien. The gesture riled the father and seemed unnecessarily irritating in light of Vincent's disclosure to Theo, at the same time, that he had no intention of renewing relationships with "the woman" (letter 351). Theo continued to send 100 francs a month; Vincent paid no board but attended to some of his debts. He argued about Theo's wanting to send an extra 50 francs to help the family out while their mother was bedridden. Vincent was reluctant on the grounds that it would seem like board, something he was determined not to contract with his father, but eventually he acquiesced.

The early months of 1884 were filled with Vincent's concern for future income and accusations directed at Theo for not trying to sell his art work (see, for example, letter 358). This two-part letter is huge; almost 8 printed pages. The argument is circuitous and tiresome but revolves around Vincent needing the monthly stipend from Theo and at the same time being upset by perceived restrictions resembling moral proclamations. He concludes with the desire not to become engaged in quarrels with "Father number two" (Theo), who can only give him money. Theo seems to have accepted this although so little of his correspondence was saved and it is difficult to judge the one-sided argument. In subsequent letters the point slowly emerged that Vincent wanted Theo to equate his allowance with payments for the art work. That none had been saleable to date was regrettable but immaterial from the artist's perspective.

In August 1884 Antoon Hermans, a retired goldsmith and antique dealer from nearby Eindhoven, commissioned six large compositions of Vincent's choosing to be used as a guide for interior decoration by the owner himself. Van Gogh provided sketches of a sower, a plower, a shepherd, the harvest, potato digging, and an ox wagon in the snow, to encompass seasons and peasant activities. Vincent later gave rudimentary lessons to two other amateur artists, Anton Kerssemakers, a reasonably well-to-do tanner and Willem van de Wakker, also of Eindhoven (letter 435b). This activity was encouraging but provided only token stipends or payments in kind.

The next month was stressful because of the Margot Begemann affair. She was a little older than Vincent, lived with her family next door, and was fond of the artist to the point of contemplating marriage. According to Vincent he could have had his way with her but did not (letter 375). The unfortunate woman was chastised by her sisters for the relationship and eventually attempted suicide with an inadequate dose of strychnine. The event occurred while Margot was on a walk with the artist, who urged her to regurgitate the poison and ran to Dr. Van der Loo in Eindhoven for an antidote. The girl recovered but was forced into separation from Vincent, who later found her to be like a Cremona violin but "rather too damaged" (letter 377).

Much time, and money, was then spent in painting peasant heads, hands, and full figures. One of the female models was rumored to be his Dulcinea or sweetheart, after Don Quixote's peasant lover, according to his friend and pupil Kerssemakers (letter 435c). She became pregnant and Vincent was suspected by the local Catholic priests and townspeople, but the artist denied it and maintained a good relationship with the girl's family (letter 423). The affairs of Margot and "Dulcinea" proved nothing more than his perceived ability with the opposite sex.

The Reverend Theodorus van Gogh died on March 26, 1885. He fell down at the entrance to his home, died instantly, and was carried into the home. Vincent's next communication to Theo (letter 397, April 1885) reveals little about his feelings but one supposes they had been ventilated during the brother's visit for the funeral. None of the children claimed an inheritance so that their mother could feel more secure. She elected to take in a boarder so Vincent removed himself to his "studio," now in a room rented from the sexton of the Roman Catholic church at Nuenen.

In spite of regular and increased "enclosures" from Theo, Vincent managed to spend more on materials and models and ran up some sizeable bills for which the brother was eventually held accountable. There was much discussion about finding outlets for the art work; a desire for his paintings to be seen and talked about; a craving for recognition. Most of the paintings from this period were significant and culminated in *The Potato Eaters,* Vincent's first real masterpiece and a "painting" rather than a "study" by his own definition. A lithograph of the ensemble was also executed in Eindhoven although some of the twenty prints were pulled on cheap paper and did not last (letter 435 c). Perusal of the letters around this period provides an exciting sequence starting with the application of finishing touches, the request for funds to cover packaging and shipment, "listening" to the paint dry, final delivery to Theo in Paris (letter 407), and then concern about its reception (see letter 424; and others).

The Nuenen period is important for Vincent's first exploration of color theory in general and his embrace of the ideas of Eugène Delacroix in particular. He asked Theo for further books on the subject (letter 428). But his hero continued to be Millet, the master of the genre. Of the new artists of France he knew but little: "There is a school – I believe – of impressionists" (letter 402).

In the fall of 1885, Vincent received assurances from Dr. Van der Loo that his mother was reasonably well and would live another ten years or so (letter 434). She actually did much better, living for another twenty-two. Vincent contemplated another move. He made a visit to Amsterdam in the company of Kerssemakers to visit museums and the two absorbed the magic of Rembrandt. But he could not bring himself to confront Uncle Cor and waited outside the establishment while his friend bought two art books (letter 435c).

Antwerp

Toward the end of November 1885 Vincent took up his new residence in Antwerp, a little room over the shop of a dealer in artist's pigments. His initial reactions to the Belgium port were positive and full of interest. The artist spent some time in the dance halls drinking beer modestly and enjoying the people vicariously. He was amused to be mistaken for a sailor (letter 437), and it is worth noting that on another occasion Vincent was taken for a manual worker by one of his doctors (letter 442, December 28, 1885). His mother is supposed to have said that he was the most robust of all of her children and Johanna van Gogh-Bonger[1] thought that her husband, Theo, looked more frail than Vincent the first time she saw them together in Paris in 1890 – all paradoxical observations in view of his illness.

Vincent's health in Antwerp was compromised by severe dental problems, which were fixed at considerable expense, and gastrointestinal problems (letter 448 and others), which were exacerbated by malnutrition and excessive smoking (see, for example, letter 449). (Illnesses are discussed in detail in Chapter 3.) He experienced periods of faintness and fever and complained of self-inflicted overwork (letter 441). Apart from this expression of underlying illness, it is clear that he had enjoyed better meals in

Nuenen and that he was again prone to ignore food in favor of art supplies while in Antwerp.

Tralbaut[3] reproduced two pages of Vincent's sketchbook from this period. On one there is a notation to indicate that he had two consultations with "A. Cavenaile" (sic) [Dr. H.A. Cavenaille] and received castor oil. Tralbaut supposed that this was for Vincent's stomach trouble but castor oil is a powerful laxative and is given for constipation. The other reads: alum, 20c – pint or 1/2 – time to time – 10 h [10 a.m.] Sitz-bath – Stuyvenberg. These notations apparently refer to alum irrigations [of the urethra] and Sitz-baths at the Stuyvenberg Hospital. According to relatives of Dr. Cavenaille, the doctor treated Vincent for syphilis, and Vincent did a portrait in lieu of payment. There is also an indication from one of the sketchbooks[8] that Vincent consulted some of the medical literature of the day, by Drs. Antoine Jozan and Joseph Capuron, on maladies of the urinary tract.

In January 1886 Van Gogh presented some examples of past work and was admitted to the Academy of Fine Arts by the director, Charles Verlat (letter 445). He took instruction from Eugeen Siberdt in drawing from the live model, Franz Vinck on drawing ornaments, and Piet van Havermaet in rendering antique objects. Verlat also provided critiques. In addition, Vincent joined some private clubs of artists to gain access to models at night. The Antwerp period was a productive learning experience. However, Vincent was argumentative and not particularly well received by instructors, and he tended to incite fellow students into more experimental approaches. Evaluation of his work was mixed; few tears were shed by the instructors after his departure.

Vincent was critical about, but not unhappy with, his formal education in Antwerp. Moreover, he soon became interested in the prospect of joining Cormon's studio in Paris even though he thought the operation was probably not that much different (letter 456). He started to persuade Theo towards that goal. In the meantime his mother had decided to leave Nuenen for Breda in March 1886, and Theo hoped that his older brother might assist with the moving. The artist felt too alienated by his sisters to visit Nuenen, anticipated that he would be too sick to be useful, and frankly did not want to make the effort. Theo had even suggested an interim in the Noord Brabant so that he would have time to assume a larger apartment in Paris. Instead, Vincent arrived unannounced in March. The last allowance had covered the train trip but much later he admitted not paying bills in Antwerp, and consequently losing some property (letter 496).

44

Paris

Much had changed in ten years and now he was in Paris as an artist rather than a clerk. This Paris sojourn would last almost two years. It was a period of discovery and familiarization with the art of the Impressionists, a time to make lasting friendships with some of them, an opportunity for further artistic development, and a savage test of Theo's patience. Theo was now the manager of a branch of Boussod & Valadon, formerly Goupil, in Montmartre. His more liberal views were suffered by the management, and he attained a position of informal leadership among the Impressionists. The importance of all of these features for the development of both brothers is clear but the amount of available information[9] does not match the interest. The explanation is simple; Vincent had little occasion to write except during Theo's business and holiday trips.

There was a communication to Horace Livens, an English artist whom Vincent first met in Antwerp, in which we find early references to Monet and Degas and an indication that Vincent spent three or four months in Cormon's studio (letter 459a). He met Émile Bernard and Henri de Toulouse-Lautrec at that liberal studio but otherwise found the experience unexciting. Fortunately we can glean from later letters, which reminisced about the Paris stint, some of the flavor of the time and the way his art work was influenced. He drew with the dexterity of hand that comes from talent plus constant practice (see, for example, Figure 2.5). Contrary to popular belief, changes in palette after coming to the French capital were far greater than those subsequently associated with the move to the south. Further discussion of temporal and regional aspects of van Gogh's "high yellow" palette is included in chapter 8.

Van Gogh turned away from the somber tones of Nuenen epitomized by *The Potato Eaters* to a brighter, livelier color scheme in Paris. In part, the change was a natural evolution resulting from further exploration of color theory, but now there was a feast of examples from current French artists and their example provided the requisite confidence-builder. The two leaders of impressionism were Claude Monet, who was already reasonably successful, and the paternal Camille Pissarro, who taught, influenced, and encouraged so many others, but was never very successful during his lifetime, and forever felt the financial struggle. Vincent also admired the contemporary accomplishments of Edgar Degas and the past advances made by Édouard Manet, but he felt more at ease with the younger artists such as Henri de Toulouse-Lautrec, Paul Signac, Georges Seurat, Armand Guillaumin, Louis Anquetin, and Émile Bernard, members of the so-called Postimpressionists.

Figure 2.5. Vincent van Gogh: *The Moulin de la Galette* 1887, Paris, ink and black chalk on laid paper, 53 × 39 cm (20.9 × 15.4″), The Phillips Collection, Washington, D.C.

Vincent had a few adventures in pointillism and the influence of Signac is apparent in his espousal of fairly broad strokes of color. But he did not have the temperament for the finer dots adopted by Seurat. This style was embraced then abandoned within the stretch of a few canvases, although Vincent later recognized Seurat as the leader of the "Petit Boulevard." This was his innovative name for the less established artists including himself, Bernard, Anquetin, Gauguin, and Lautrec who showed at the Café du Tambourin, on the boulevard de Clichy, compared with Monet, Sisley, Renoir, Degas, and others that Theo showed on the "Grand Boulevard," i.e., 19 boulevard Montmartre.

Van Gogh was also intrigued by the compositions, brave color schemes, and strong strokes of Adolphe Monticelli, who had spent time in Paris and then resided for most of his productive career in Marseilles. The work of Monticelli was very influential on Vincent van Gogh; there were also life-style similarities that developed in the South. Vincent felt that Monticelli had come in a direct professional line from Eugène Delacroix, and later he referred to some of his own work as evolving into a "metaphysical philosophy of color à la Monticelli" (letter 503).

In the summer of 1886 Theo was embroiled in a relationship with the mysterious "S." Vincent offered advice about the mutual unsuitability of his brother and S. While Theo was in Holland, unsuccessfully seeking support from their uncles for his own gallery, the trio of Vincent, Andries Bonger (Theo's brother-in-law-to-be), and S., enjoyed a roller coaster existence in his apartment. The woman had alternating periods of lightheartedness and serious derangement. Theo clearly wished to extricate himself. In an amazing tour de force of social welfare Vincent offered to "take S. off [Theo's] hands ... even agreeing to a marriage de raison" (letter 460). It is generally assumed that S. is not Segatori (see below) because Vincent openly recognized her the following year.

Vincent became a regular at the Café du Tambourin, where he was introduced to the charms of absinthe by Toulouse-Lautrec. The Lautrec pastel of Vincent sitting with a glass of the liqueur is a sensitive portrait from that time. The manager, Agostina Segatori, was an experienced model and she posed at least twice for Vincent; *Woman at a Table in the Café du Tambourin* and *The Italian Woman*. It has been suggested that "La Segatori" had an affair with Vincent, that she was the subject of one of his "impossible, and not very seemly, love affairs" which left him "damaged and shamed" (letter W1). Vincent had an arrangement to eat at her Café in return for canvases as decorations, which gradually accumulated on the walls. He reclaimed them in the summer of 1887, although it was a source of contention (letter 461). Even then he felt affection for her and hoped she still had some for him (letter 462).

According to the memoir of Johanna van Gogh-Bonger,[1] and less accessible letters by Theo to his sisters,[4] the friendship between the brothers became quite strained. Theo complained that his "home life [was] almost unbearable," and that his brother seemed "as if he were two persons: one marvelously gifted ... the other egoistic and hard-headed." Separation within Paris was impractical and Theo later admitted that his *own illness* made him less than tolerant of Vincent. Andries Bonger confirmed that Theo looked frightfully ill as early as June 23, 1886 and not much better by February 18, 1887 (letter 462a).

Vincent also developed a mutual regard for Julien Tanguy, a placid paternal figure who narrowly escaped death during the Franco-Prussian War, and had since been a dealer in artist's pigments. His shop was close to Vincent's residence. "Père" Tanguy was the subject of two large canvases. He carried Vincent's work for several years and sold one for which Vincent was credited 20 francs (letter 506). While in Paris, and later in the South, Vincent continued to order some of his pigments from Tanguy out of loyalty (letter 503). Tanguy's wife was less sympathetic with the financial plight of artists and probably initiated the dunning that Vincent subsequently received. In a most humorous bit of creative accounting Vincent supposed he could send a return bill to Tanguy to include two items, "money Tanguy has made on [sending me] paints" and "friendship," each worth 50 francs (letter 510). Nonetheless there followed many positive references to Père Tanguy and the old dealer was a pallbearer at Vincent's funeral.

Vincent's introduction to things Japanese, *Japonaiseries* as they were called, was significant. In Japan cheap prints were used to wrap oriental crockery for shipment to Europe and these were the first Japanese works on paper embraced by the young artists of Paris. An interest in the finer wood-block prints followed. Vincent was not alone in taking stylistic succor from the Orient. The development of cloisonism, the use of simplistic, clearly-outlined forms in flat colors, by Émile Bernard and Paul Gauguin was clearly influenced by Japanese compositions. Vincent frequented the shop of Siegfried Bing, who reputedly had ten thousand prints in the attic; van Gogh took about ninety francs worth on consignment but didn't sell many. However, they were images of lasting influence which were later incorporated into several paintings. The meanings behind the subjects were reinterpreted as Vincent became more familiar with the realities of Japanese culture (letter 505 et al.).

As early as October 1886, Vincent expressed a desire to visit the south of France (letter 459a), and Toulouse-Lautrec is credited with recommending Arles. Vincent had a general feeling that the exotic aspects of warmer climes might be the "Japanese" of France, and many months later (from St. Rémy, 10 September 1889) reminisced that he "came to the South ... to see a different light ... [to get] a better idea of the Japanese way of feeling ... to see this stronger sun ... [and to experience those] colors of the prism [which are] veiled in the mist of the North" (letter 605). Several other reasons have been offered for his departure from Paris not least of which that both Theo van Gogh and Andries Bonger thought Vincent was disruptive; they ventilated their feelings in letters to relatives and their sisters thought they were more than patient. However, it is not clear that Vincent perceived this tension and there is nothing in his letters along those lines.

The growing attachment to Adolphe Monticelli was certainly a factor. Vincent wanted to explore the haunts of this artist, who died in Marseilles in 1886. To his sister [in 1888] he wrote, "I think of Monticelli terribly often here [Arles]. He was a strong man – a little cracked, or rather very much so ... always harassed by poverty ... an extremely refined taste as a colorist ... for myself, I am sure that I am continuing his work here as if I were his son or his brother" (letter W8). The next year, Vincent described himself as Monticelli's heir. Towards the end of his life, after Albert Aurier published the first major review of van Gogh's work, Vincent lauded the ability of Monticelli and warmly acknowledged Monticelli's influence upon his own painting (letter 626a). Vincent always planned to end up in Marseilles, but never set foot in the port city.

At various times Vincent urged Theo to corner the market on Monticelli's paintings and the van Goghs eventually acquired six Monticellis.[10,11] Three of these were reproduced as lithographic copies by Auguste Lauzet in a large-format book that was edited by the poet Paul Guigou and published by Boussod & Valadon in 1890.[12] In this and other ventures they were potentially interested in a collaboration with the Scottish art dealer Alexander Reid.

Van Gogh had painted Reid's portrait while he was working a few months in Paris with Theo, and Reid exchanged a Monticelli for a van Gogh still life. During 1887 Reid and the van Gogh brothers frequented the gallery of Joseph Delarbeyrette, who was Monticelli's main outlet in Paris. After reaching Arles, Vincent wondered whether their relationship with Reid, concerning Monticelli as well as the Impressionists, would be one of cooperation or competition. It turned out to be neither. Reid eventually went to Marseilles, gradually acquired Monticelli's pictures for his own gallery, became a significant dealer in French art, but unfortunately did not carry van Gogh's work until the 1920's. Vincent later declared that he had been unduly rough with Reid (letter 597).

The Paris winters of 1886 and 1887 were unusually cold. Perhaps Vincent was ready to try a warmer locality but I think that the primary reason he moved to the south of France was a lack of acceptance in Paris. There are several anecdotes about Vincent's attempts at showing his works in small galleries, cafés, and during friendly gatherings. Suzanne Valadon[11] described one such episode in connection with visits to Toulouse-Lautrec's studio. Vincent's timing was usually inappropriate and his reception was not particularly warm, even among colleagues whom he perceived as otherwise sympathetic and talented. He resurrected the idea of a co-operative, a place for the struggling artist like himself, for those who were sufficiently unencumbered to travel but prepared to live "like a monk." In several letters from Arles he endeavored to sell the concept of a cooperative studio-dwelling in the South. In another (letter 531) he invoked a similar setup in the Drenthe (under the proposed leadership of Eugène Boch) so that participants "could change places – thus being sometimes in the North." The "Yellow House" would prove to be the short-lived culmination of these dreams. I believe that Vincent's evangelism had been completely transferred from religion to art by the time he left Paris for the South.

Vincent decorated Theo's apartment with his canvases, paid a last minute visit to the studio of Seurat, packed his clothes and materials, including the perspective frame, and jumped on the overnight train to the South (the Paris-Lyon-Marseilles Express). He landed February 20, 1888, in Arles, a town dating from Roman times, on the Rhone River. He was slightly irked to find two feet of snow on the ground, and still more falling (letter 463). He took a room within walking distance of the station.

Arles

His first address[*] was the Hôtel-Restaurant Carrel, 30 rue Cavalerie, whose dining room was later the subject of a painting.[13] On May 1, 1888 he rented a house, actually the right wing of a complex, a few blocks away at 2 place Lamartine; it contained four rooms, two up, two down, and was painted yellow outside. For a short time the following year he also rented two other rooms in the complex in anticipation of visitors (letter 570). This building was immortalized as *The Yellow House* in a drawing, a water color, and an oil painting. It had been vacant for some time and was in need of repairs and improvements; there was running water, but the toilet was in the hotel next door (letter 480). Vincent did not move in immediately but took his meals at Restaurant Venissac, 28 place Lamartine, and continued to live at the Carrel until litigation with the proprietor (it was resolved by the magistrate

---

[*] The chronology of the Arles period is well documented in the text of the catalog edited by Pickvance,[13] wherein most of the highlights are registered to the exact day.

in partial favor of Vincent) forced him to move on May 7, 1888, to Café de la Gare, subject of *The Night Café*, at 30 place Lamartine. He did not sleep in the Yellow House until September 17, 1888, after helping with the cost of repainting and paying for the installation of illumination gas. Place Lamartine is named for the poet Alphonse de Lamartine who, as politician, had been instrumental in bringing the railroad to Arles. These addresses were all clustered in the north end of town; unfortunately, all of these buildings have since been destroyed.

In the early months at Arles Vincent learned that Theo might be sent to the United States on behalf of Boussod & Valadon. Later, in a moment of optimism, Vincent wondered whether B. & V. might reinstate him and allow him to accompany his brother (letter 492). It was not to be, but Vincent's idea did prompt Theo to again explore the possibility of opening his own establishment in Paris. He hoped for aid from his uncles. Vincent also wondered about his own prospects of promoting exhibitions and opening a dealership in Marseilles; he urged Theo to pull Tersteeg into the enterprise by promoting the Impressionists in Holland and England (letter 465). All of these ideas are attractive in retrospect but were ahead of their time and came to naught.

During 1888, there were numerous clues from correspondence that both of the van Gogh brothers were enjoying anything but the best of health. Doctors Louis Rivet and David Gruby of Paris were frequently mentioned, and it is clear that Vincent himself identified his stay in Paris as the start of excessive drinking and smoking. There was a discourse on the doubtful value of potassium iodide as a tonic together with an assessment of the relative merits of Rivet and Gruby as general practitioners (letter 489). The level of discussion between Vincent and Theo suggests a reasonably high degree of observational skills on matters medical. In a cumulative fashion, Vincent's references elicit our confidence in his ability to report his symptoms as a patient, albeit in lay terms.

Vincent was also friendly with an erstwhile medical student from Denmark, Christian Mourier-Petersen, who was sufficiently disillusioned with medical practice to claim that "It's the doctors that kill people." He was an amateur artist who had come south for "a nervous disorder, which had been brought on by the strain of the examinations" (letter 490). Vincent's friend moved to Paris later and boarded with Theo for several months. In many ways Mourier-Petersen strikes me as the epitome of the companions with similar and sympathetic ailments who managed to maintain a lasting affinity for Vincent van Gogh.

Van Gogh's early letter from Arles to Toulouse-Lautrec went unanswered. However, there is no evidence to indicate a lessening of mutual esteem. Lautrec stoutly defended the art of van Gogh in Paris and later in Brussels,

Figure 2.6. Vincent van Gogh: *Garden with Weeping Tree*, 1888, Arles, pencil, quill and reed pen with brown and black ink on wove paper, 24.46 × 32.08 cm (9.63 × 12.63"), Courtesy of The Menil Collection, Houston, Texas

and they had a brief but warm renewal of friendship in 1890. Van Gogh also corresponded with the Australian artist John Russell, who was financially secure and received numerous requests from Vincent to buy paintings from Gauguin and Bernard. Russell eventually assembled a good collection which included some of Vincent's art.

Van Gogh's letters to Émile Bernard from this period were extensive and are collected in a separate section of the *Complete Letters*.[14] It was to Bernard in March of 1888 that Vincent complained that life in Arles was not as cheap as he had hoped (letter B2). Bernard was also a poet and sent examples for Vincent's comments (letter B4). Letters by Vincent to Theo and to Bernard were full of descriptions of current paintings. Occasionally they were quickly sketched between paragraphs, and then annotated; for example, "Here is a new subject, a corner of a garden with clipped shrubs and weeping tree, and in the background some clumps of oleanders, and the lawn just cut with long trails of hay drying in the sun, and a little corner of blue sky at the top" (letter 508, and see Figure 2.6 for the original drawing).

Vincent felt that Bernard, Gauguin, and himself should have gone to the same place, specifically Arles (letter B7). One may speculate that Émile Bernard, who was fifteen years younger than Vincent, more willing to exchange work, and a less-domineering personality than Paul Gauguin, might have made a better colleague in the Yellow House. There were some amusing exchanges in letters. Although barely risqué by today's standards, they suggest that Vincent was even prepared to assume the role of experienced advisor for a young Bernard preoccupied with brothels (letters B4, B7, B8, B14). There were several suggestions that Émile might join him, especially after the already grudging support by Bernard's father was further threatened. Vincent had had some unpleasant encounters in Paris, wherein he supported artistic ideals against the senior Bernard (shades of arguments with father van Gogh). The exchanges were conducted with such vehemence that van Gogh was banished from the household.

The American Dodge MacKnight, who spent some time in Fontville, about six miles from Arles, and the Belgian Eugène Boch were also invited to reside in the Yellow House. Mourier-Petersen felt he had spent enough time in the South, and Bernard couldn't make up his mind whether he wanted to serve his military obligation. So none of the four elected to join Vincent during what proved to be a rather short tenure on the Yellow House.

Anton Mauve died in 1888. After receiving the obituary notice in late March Vincent was moved to dedicate his just finished painting *Pink Peach Trees*, to his first mentor. It was one of his best landscapes to date and was inscribed "Souvenir de Mauve." It was signed "Vincent," not "Vincent and Theo" as he claimed in letters 472 and W3, although Theo was asked to send it to the widow, their cousin Auriette (Jet) Mauve, from both of them. It was well received by Jet Mauve who spoke also of better days, when relationships had been so good (letter 562). He was obviously feeling nostalgic for Holland because he dedicated a version of *The Langlois Bridge* to H. G. Tersteeg, and a still life to the Dutch painter George Breitner, with whom he had worked in The Hague in 1882, and he gave a picture to his sister Wil in early April. Tersteeg was still unable to move impressionist or postimpressionist paintings in The Hague. For example, in midyear 1888, Theo sent him a consignment of ten pictures; they included major works by Degas and Gauguin and a Paris picture by Vincent (letter W4). They all came back next month, not one was sold.

Vincent continued his avid pursuit of good literature;[15] examples are collected in Table 1. He absorbed the culture of the South by reading Alphonse Daudet's *Tartarin sur les Alpes* and made several allusions to the hero of that story (see, for example, letter 469) but later claimed that he saw little of the "Southern gaiety that Daudet talks about so much" (letter 502). He made several visits to the abandoned Abbey of Montmajour and compared the surrounds to "Paradou" in Zola's *La Faute de l'Abbé Mouret* (The Sin of Father Mouret) (letter 506). It is paradoxical to note, months later, that the last book Vincent read in Arles was Balzac's *Le Médecin de Campagne* (The Country Doctor) in which van Gogh particularly enjoyed a female character who was "not mad but too sensitive, which is very attractive"

Table 1. Books read by Vincent van Gogh*

Locality (number)	Number per year	Examples (Author, *title*)
London (12)	8	Ernest Renan, *Jésus.* Victor Hugo, *Les Misérables.*
Paris (18)	16	Jules Breton, *Les Champs et la Mer.*
Ramsgate et al. (11)	16	John Bunyan, *Pilgrim's Progress.*
Dordrecht (11)	37	Collin de Plancy, *Légendes des Artistés.*
Amsterdam (19)	17	Thomas à Kempis, *Imitation of Christ.* Jules Michelet, *Histoire de la Révolution.*
Etten, Borinage, Brussels (9)	3	Harriet Beecher Stowe, *Uncle Tom's Cabin.* J.C. Lavater & F.J. Gall, *Physiognomy and Phrenology.*
Etten (10)	14	Jules Michelet, *Du Prêtre, de la Femme, de la Famille.* Charlotte Brontë, *Jane Eyre.*
The Hague (34)	20	Émile Zola, *Nana; La Faute de l'Abbé Mouret; Son Excellence Eugène Rougon; Au Bonheur des Dames.*
Drenthe (3)	13	Thomas Carlyle, *Oliver Cromwell's Letters and Speeches: with elucidations; On Heroes, Hero-Worship, and the Heroic in History.*
Nuenen (15)	8	Gustave Flaubert, *Madame Bovary.* Alphonse Daudet, *Histoire de Mon Livre.* Alfred Sensier *[on Millet, Rousseau, Michel].*
Antwerp (5)	19	Émile Zola, *L'Oeuvre.*
Paris (2)	1	Guy de Maupassant, *Bel-Ami.*
Arles (23)	19	Guy de Maupassant, *Pierre et Jean.* Alphonse Daudet, *Tartarin sur les Alpes.* Victor Hugo, *L'Année Terrible.* Honoré de Balzac, *Le Médecin de Campagne.* Ernest Renan, *L'Antéchrist.* Charles Dickens, *Christmas Books.* Edmond & Jules de Goncourt, *Les Frères Zemganno.* Gustave Flaubert, *Bouvard et Pécuchet.*
St. Rémy (12)	12	Guy de Maupassant, *Fort comme la Mort.* William Shakespeare, *Richard II* etc. Edouard Rod, *Les Sens de la Vie.*
Various and repeated (6)	?	*Bible.* Johann Wolfgang von Goethe, *Faust.* Miguel Cervantes de Saavedra, *Don Quixote.*

* The data were extracted from Barr AH. 1936, Vincent van Gogh. New York: The Museum of Modern Art, and confirmed in the *Complete Letters*. The second period in Paris is certainly underestimated because of the paucity of letters from Vincent during this period.

(letter 590). The same letter also contains the delightful observation, "They have lots of room here in the hospital, there would be enough to make studios for a score or so of painters."

The early letters from Arles are full of sanctimonious advice to Theo. A particularly long discourse along these lines is letter 489 in which Vincent eventually reminds Theo that he can still see Dr. Gruby in Paris saying "No women!" but himself thinking, "when you have to work all day with your brain, ... you've had as much as your nerves can stand." There are further and repeated warnings about the dangers of smoking, drinking, and sex. Notwithstanding best intentions for moderation it is clear that van Gogh resumed all three fairly heavily. For example, "I saw a brothel here last Sunday – not counting the other days ... I shall have to make a lot of noise as I aspire to share the glory of the immortal Tartarin de Tarascon" (letter B4). Later, the brief visit by Paul Gauguin was to do more harm than good for Vincent's sobriety.

Uncle Cent was suffering (letter 512) and died shortly thereafter (letter 516), as did the last hope of Theo's receiving substantial family support for an independent gallery. Vincent described, "an image of the man made up of memories of so long ago," and it seemed to him, "so peculiar that a man [I] once knew at such close range should have become such a stranger" (letter W5). Theo received some inheritance from Uncle Cent, but Vincent always considered the lack of major support by Uncles Cent and Cor to be regrettable and a grave error on their part. Vincent and Theo continued to converse with Tersteeg, but he was either unwilling or unable to bring much support to Vincent or artists in a similar position. Towards the end of July 1888, torn by disappointments, and in less than full health, Vincent nonetheless waxed poetic, "The more I am spent, ill, a broken pitcher, by so much more am I an artist – a creative artist – in this great renaissance of art of which we speak" (letter 514). The concept that something less than normal health was compatible with artistic production and perhaps even conducive of creative work is a recurring theme.

The letters from Arles, and later from St. Rémy, provide interesting documentation on the dates of paintings undertaken by Vincent as well as the various groups of canvases that were shipped to Paris.[13,16] They also make clear Vincent's distinction between studies and paintings. *The Sower* and *The Night Café* were finished works in the latter category (letter 534). *The Night Café* was started on the evening of September 4, 1888, (letter 518), and Vincent worked under the ambient gaslight for a further two nights, "to express the terrible passions of humanity by means of red and green" (letter 533). Vincent deemed it to be equivalent to *The Potato Eaters* and it was addressed in several subsequent communications. "In my picture of the night café I have tried to express the idea that the café is a place where one can ruin oneself, go mad, or commit a crime ... an atmosphere like a devil's furnace of pale sulfur" (letter 534). One supposes that Vincent had a few drinks on the spot and in any event he guessed that Tersteeg might say that the picture was at least "delirium tremens in full swing" compared with a gentle canvas by Sisley, which his former employer deemed to be by a slightly tipsy artist (letter 534).

In mid-September Vincent purchased a mirror expressly to do self-portraits in lieu of working from models. All of the self-portraits were supposedly done this way, i.e. the mirror image is depicted. At this stage Vincent was working at a furious pace, sometimes six and twelve hour stretches of painting followed by twelve hours of sleep (letter 537). The enterprise was partly driven by the desire to accumulate a body of work before the hoped-for visit of Gauguin. It was always Vincent's intention to assemble a significant number of paintings, often with series on the same subject. In many cases he contended that they would show better together in meaningful combinations rather than as separate entities. This was attempted in the exhibitions[17,18] in Holland in 1990.

After lengthy correspondence in which Vincent expressed degrees of hope, doubt, despair, and even resignation, Paul Gauguin finally joined him on October 23, 1888. It is generally agreed that Vincent van Gogh reached the peak of his creative powers in Arles. I prefer to call it a high plateau; there is no evidence of a decline in St. Rémy or Auvers. The interaction between the two artists was important and electric but, contrary to the popular impression derived from popular film representation, they spent only two months together and Gauguin made a hasty departure for Paris on December 25, 1888. Notwithstanding the claims of mentorship by Gauguin,[19] it is clear that Vincent had reached a consistently high level of artistic skill before Gauguin's arrival. Nonetheless, they exchanged art work and it is also apparent that the two artists had a positive influence on each other. This was more readily acknowledged by van Gogh[1] than by Gauguin.[19] In letter 626a, to Vincent's first major reviewer, Albert Aurier, van Gogh gave a warm acknowledgment of Gauguin's influence.

From all accounts, including his own, Paul Gauguin brought a degree of organization to the day-to-day affairs of the domicile. This included a certain amount of home-cooking, joint trips to the brothel to preserve "mental hygiene," and a rudimentary attempt at budgeting the stipend from Theo. There was productivity; in the eight weeks he was in Arles Gauguin produced at least sixteen pictures and van Gogh probably twice that number. Gauguin had come under desperate straits and, notwithstanding the generosity of Theo to carry the pair, they seem to have reinforced each other's discomforts about the lack of recognition and the difficulty in selling canvases. Theo tried to put Vincent at ease by deeming the whole question of money, and sale of pictures, as not important or "existing as a disease," and expressing thanks for his brother's ability to create an entourage of artists and friends for Theo (letter T3). It is somewhat paradoxical that it was Vincent, not Theo, who raised doubts about the extent of financial support extended to Gauguin, and he strove to establish and maintain an obligation; i.e. that Gauguin provide Theo with an appropriate number of paintings (letter 536). The reasons may have been complex, but it is sad to find Vincent, a year and a half later, expressing doubts about his brother's sincerity.

Van Gogh and Gauguin made an excursion to Montpellier to view the collection of Bruyas. This was a seminal encounter for both artists but differences in opinion about the relative contributions of the artists in that outstanding collection, and subsequent discussions on art and philosophy generally, degenerated into abrasive comments from both sides. That their relationship should be brought to such a pitch of heated argument was but one more example in Vincent's list which included: his father, H.G. Ter-steeg, Rev. Stricker, Anton Mauve, Alexander Reid, Émile Bernard's father, and eventually Dr. Gachet.

Alarming but poorly documented events included Vincent's throwing a glass of absinthe at Gauguin's head and, some days later, confronting him in the street, with or without a cut-throat razor, according to skimpy and contradictory versions due to Gauguin and, indirectly, Bernard. The threat was averted by a strong directive from Gauguin who nonetheless remained sufficiently enervated to seek a hotel room for the night.[19] During the hours that followed Vincent cut off part of his left ear lobe, wrapped the dismembered piece, and presented it to a prostitute. Vincent returned to his own bedroom and was discovered the next day by the police and, thanks to the intervention of his friend, the postal worker Joseph Roulin, was eventually taken to the Hôtel Dieu, the hospital in Arles, where he was attended by Félix Rey. Rey was a young intern still in training; the director of the hospital was Dr. Urpar but he is not mentioned by Vincent in the extant letters. The ear-cutting incident started on the night of December 23 and played over into Christmas eve. (See Chapter 9 for more discussion.) It received a paragraph in the newspaper, Forum Républicain, the following week.

Theo van Gogh had planned to become engaged to Johanna Bonger in 1886, but it was not until December 1888 that his life became sufficiently unencumbered to propose. It was therefore most unfortunate timing to receive a telegram from Gauguin, on the evening of December 23, informing him of Vincent's state of "high fever combined with delirium."[3] He took the night train and arrived the next day in Arles. After conferring with doctors at the hospital, Joseph Roulin, and also the pastor of the Reformed

Protestant church, Reverend Frédéric Salles, Theo returned to Paris on the 26th. His reasons for not staying longer have not been determined. Gauguin also took the train to Paris, most likely on Christmas day. It was left to Rey, Roulin, and Salles to keep Theo informed of developments.

The early prognosis was poor. (This subject is discussed in further detail in Chapters 3 through 6.) After Mrs. Roulin visited on the 27th, Vincent had a second attack, and the following day Mr. Roulin was not able to see him because Vincent was completely unable to communicate, i.e. suffering from aphasia. And then on the last day of December, to the pleasant surprise of doctors and friends, the patient made a rapid recovery to the extent that the Rev. Salles could report that he found him "calm, in a state which revealed nothing abnormal." By the first week of January Vincent was moving around the hospital and conversing freely with Roulin and others, and even cautioning Theo not to alarm his mother and sister unduly (letter 569). On January 7, he returned to the Yellow House and that day declared to his mother and sister that "there is a chance that there will be nothing the matter with me for a long time to come" (letter 569a).

On January 9, Vincent told Theo that he was suffering from insomnia and, without consultation with the doctor, was treating himself with massive amounts of camphor (letter 570). It was an exaggerated application after reading Raspail's book of home remedies (letter 576), and we shall return to this in chapters 4, 5, and 7. The next week Vincent started to analyze his illness and reported that Dr. Rey says, "that I really must feed myself." It is appropriate to mention here, and to analyze further in chapter 3, that Vincent's reaction was a foreshadowing of a tendency to noncompliance and the seat of an important exacerbation factor. "I took the liberty of saying to Mr. Rey, that ... by pure chance or misunderstanding ... I had had to keep a strict fast for a week ... [and] had [he] seen many madmen ... fairly quiet and able to work [like me]" (letter 571). There is good evidence that malnutrition, and especially fasting, were highly detrimental to Vincent's condition.

Van Gogh suffered from "unbearable hallucinations" during attacks, and had nightmares in the intervening periods; he attributed some relief to taking potassium bromide (letter 574) and felt that during major crises he was "out of [his] wits" (letter 576) and "didn't in the least know what [he] said, what [he] wanted, and what [he] did" (letter W11). There is no indication that Vincent's self-mutilation was a conscious act; he had no recollection of the events that evening (see also chapter 9). When he returned to the brothel, where he had left his unwholesome present, it was as much a mission of discovery as an attempt to make amends (letter 576). Although the feeling of past sickness developed slowly after leaving the hospital (letter 576), he was clearly shaken by events. Nonetheless he resumed painting with intensity and his canvases included a portrait of Dr. Rey and several featuring Mrs. Roulin.

During the first week of February, Vincent suffered a relapse in which he showed renewed signs of mental distress including feelings of persecution. He was again taken to the hospital for a total of ten days during which time the Rev. Salles found him at first, February 7, 1889, to be "hiding himself in absolute silence ... weeping without uttering a single word ... and refusing all food"[4] but on the day of discharge, February 17, Vincent wrote with complete lucidity (letter 577). We find that attending physicians, friends such as Roulin and Salles, and even the artist himself were surprised and encouraged by the rapid recoveries after each crisis. Likewise, the lucidity with which the patient comprehended and wrote letters, discussed his condition with physicians, weighed the possibilities for the future, and maintained the quality of his art work, are all evident. Rapidity of recovery and intervening lucidity are pivotal findings for the diagnosis of Vincent's underlying disease (and see chapter 5).

Vincent returned to work at the Yellow House but was resigned to the possibility that a further medical crisis might lead to his confinement in a mental asylum (letter 577). As a precaution he continued to eat and sleep at the hospital. He perceived that the locals had a "superstitious" view, which was directed at his work as well as his person. Their attitude grew into collective derision, especially by the youngsters in the neighborhood, and his worst fears were realized by a petition seeking intervention by the mayor. Their claims of excessive drinking and unruly behavior by van Gogh were foremost but, as Salles supposed, "his crazy act which necessitated first hospitalization" was a persistent alienation and would necessitate Vincent finding another work place.[4] Action on the petition resulted in his involuntary confinement at the Arles Hospital, where Salles found him on March 18 to be "entirely lucid, and with a complete awareness of his [own] position."[4] These events were summarized by the artist on March 19, after almost a month's silence. He felt "absolutely calm at the present moment," regretted being described as "a man not fit to be at liberty," but did not find the mayor or the commissioner of police "unfriendly." He was more concerned about provision of a studio and the hospital's pestering him about his drinking and smoking habits (letter 579).

On March 23, Vincent was visited by Paul Signac, who was interviewed about it many years later by Gustave Cocquiot.[20] Signac was a good friend and took van Gogh out for relaxation, visited his old quarters in Arles, admired his paintings, and reminisced about better times. Toward the end of the evening he had to restrain Vincent, who "wanted to drink about a quart of essence of turpentine from the bottle." As Signac said, "It was high time to return to the asylum." Elsewhere[21] I have made the chemical connection between turpentine, camphor and other terpenes and Vincent's apparent affinity for them (and see chapters 3 through 8).

The report given by Signac to Theo does not mention the turpentine incident, but his opening remarks, "I found your brother in perfect health, physically and mentally" were surely an attempt to sustain hopes, and his omission of other events from this letter should not be interpreted as negation. Moreover, he went on to say that Dr. Rey, "is of the opinion that if he [Vincent] should lead a very methodical life, eating and drinking normally and at regular hours, there would be every chance that the terrible crises would not repeat themselves at all" (letter 581a). And Vincent himself confirmed this, "M. Rey says that instead of eating enough and at regular times, I kept myself going on coffee and alcohol. *I admit all that*" (my italics, letter 581).

Vincent was pleased to find out that "the real neighbors, those whom I knew, were not among the petitioners" but he acknowledged the necessity for Mr. Salles to look for "an apartment in another part of town" (letter 582). Vincent felt that he was "absent-minded and could not direct [his] own life right now" (letter 586). Shortly thereafter he raised the curious option of joining the French Foreign Legion, notwithstanding doubts that they would accept him for health reasons (letters 588 & 589). Theo replied that he disapproved (letter T6), and the idea died a natural death.

The prospects of going to an asylum in Aix, Marseilles, or St. Rémy were raised, discussed in terms of expense, and weighed for protection versus relative freedom to continue painting. The elected plan was to go to the St. Rémy asylum *for three months* (my italics, letter W11). In no small part the choice of place and event were determined by Vincent's desire to commit himself to an institution in contradistinction to being detained after police action, as had happened in Arles. It turned out that self-admission was not sufficient; Theo had to make the formal request and it is worth noting that Vincent's release the following year was also contingent upon his younger brother's formal request. Theo's letter to the Saint Paul de Mausole Asylum at St. Rémy is reproduced in Tralbaut's account.[3] It includes a request that his brother be allowed to paint outside the establishment and that he be allowed at least half a litre of wine with his meals. On May 8, 1889, Vincent made the short journey by train to St. Rémy in the company of the Reverend Salles.

St. Rémy

They were met by the director, Dr. Théophile Peyron who noted in the register the following day that Vincent had been treated in the hospital at

Arles and was suffering from "acute mania with hallucinations of sight and hearing, which have caused him to mutilate himself by cutting off his right ear [*sic*, it was part of the left ear]." The voluntary nature of admission was recorded together with the opinion that "M. van Gogh is subject to epileptic fits at very infrequent intervals." Vincent had raised the possibility of epilepsy in Arles (letter 589), although there is no direct evidence that the doctors there had arrived at such a firm diagnosis (see chapters 6 and 7). Dr. Peyron noted that the patient himself referred to cases of epilepsy on his mother's side of the family. In Vincent's first letter from St. Rémy he said,"the doctor here is inclined to consider what I have had [was] some sort of epileptic attack" (letter 591). It is highly probable that the director at St. Rémy embraced Vincent's personal diagnosis rather than formulating his own.

The benefits derived by Vincent from internment stemmed from a structured environment, regular meals, decreased alcohol intake, and ingestion of potassium bromide as a calmative, assuming that the beneficial effects of bromides at Arles prompted continuation at St. Rémy. After a

Figure 2.7. Vincent van Gogh: *Wheat Field with Rising Sun*, 1889, St. Rémy, black chalk, quill and reed pen, black and brown ink on toned paper, 47 × 62 cm (18.5 × 24.4"), Staatliche Graphische Sammlung, München

period of orientation and assessment he was given an extra room for painting. (One field of view from Vincent's room is shown in Figure 2.7.) The patient was eventually allowed the latitude of selecting motifs outside the walls of the institution although he was accompanied by an attendant on these jaunts. His painting constituted what we would now call "work therapy." Dr. Peyron's permission to paint may have been more a concession than a prescription; see chapters 6 and 7. Vincent's observation that the other inmates "do absolutely nothing" (letter 592), suggests that this was not active institutional policy but rather an accommodation of Theo's original request. There are numerous references within the correspondence to Vincent's becoming engrossed in his work and being gradually distracted from fears of returning ill health. He was interviewed by Dr. Peyron at regular intervals and also received two hour baths, twice a week (letter 592). "Water therapy" was in vogue.

However, the attacks returned and Vincent was subject to four more major crises while at St. Rémy (see Figure 2.8). It is of interest to note that three of the four were closely associated with preceding visits to Arles. Vincent, in the company of an attendant, visited the Ginoux family at Café de la Gare and his old charwoman from the Yellow House on July 7, 1889; he had an attack on July 16 and he was severely debilitated for about 45 days. On November 17, Vincent referred to a recent visit to Arles for two days to see the Ginoux family again and also old neighbors, who were "very friendly, and even welcomed me" and he was curious to see "if this journey will provoke another attack" (letter 614). The next crisis did not come until December 24, and the delay in this one case was sufficient for Hulsker[4] and some other commentators to dismiss the connection in the other three. On January 21, just two days after another visit to Arles, a further crisis ensued and lasted seven days (see footnote to letter 624). And then perhaps the most debilitating, and certainly the longest, period of illness started on February 23 (see footnote to letter 628) and did not resolve until about

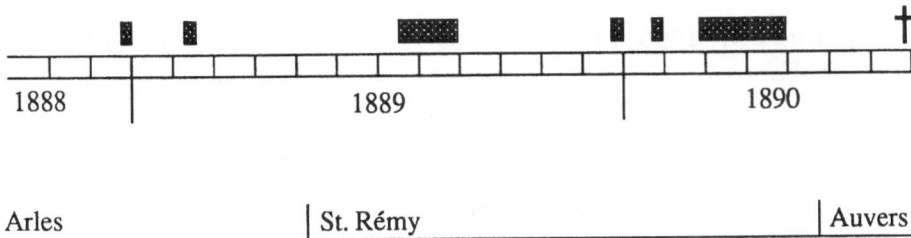

Figure 2.8. Van Gogh's crises. The center track shows calendar years marked in months. The periods of debilitating illness are depicted as upper track stippled bars. Locations are indicated on the lower track.

April 29 (letter 629), a total of 65 days. This one actually started in Arles, the day after he arrived for a scheduled two day visit, and Dr. Peyron had to send a carriage to bring Vincent back to the Asylum. It is worth looking more closely at the exceptional case, the December 1889 attack.

I quote from two letters by Theo to Vincent; first, "if you know that it is dangerous for you to have colors [artist's oil based paints, thinned with turpentine] near you why don't you clear them away for a time and make drawings?" (letter T 23; January 3, 1890), and second, "he [Dr. Peyron] gave me to understand that it was dangerous for you to go on painting, as the colors were poison to you" (letter T 24; January 8, 1890). Dr. Peyron thought that Vincent was trying to poison himself. I do not think so, it would be one of the most ineffectual methods, and see my article[21] and chapters 3 and 4 suggesting that Vincent developed a pica for terpenes. It is worth quoting again from Dr. Peyron's entry in the register at St. Rémy, "the patient ... experienced during his stay in the institution several attacks with a duration of two weeks to a month [*sic* - actually 45, 7, 7, and 65 days]; during these attacks the patient is the victim of terrible anxieties, and he has repeatedly tried to poison himself either by swallowing the colors which he used in painting or by drinking the petrol ... [used in] lamps. In between the attacks the patient is completely calm and lucid; he then abandons himself with passion to his painting." Tralbaut[3] claims to have confirmed the attempts at paint-eating from an interview with one of the attendants, J-F. Poulet.

There is an additional suggestion, perhaps indirectly related to the pica, on the earlier July–August crisis. In letter 601 (August 22, 1889) Vincent admitted that he found it, "very difficult to write, my head is so disordered. ... terribly distressed because the attacks have come back ... For many days my mind has been absolutely wandering, as in Arles, quite as much if not worse." According to Hulsker[4] this letter was so depressing to the family that they excluded part of it from *The Complete Letters*. In particular, "It seems that I [Vincent] pick up des saletés [dirt] and eat it, although my memories of these bad moments are vague, and it seems to me there is something shady in this story." We will return in chapters 4 and 5 to the nature of the pica, and to turpentine, camphor, and the terpenes of absinthe as possible influences on van Gogh's medical crises.

On January 31, 1890, Theo wrote to Vincent about the birth of his son, "a beautiful boy who cries a good deal, but who looks healthy" (letter T27). He was named Vincent Willem van Gogh (1890–1978), after his uncle. The nephew would later help in finishing the translation and assembly of *The Complete Letters* and therein he added a short memoir on his mother Johanna van Gogh-Bonger. In the van Gogh literature he is often referred to as "Vincent the engineer." Vincent replied to this good news during the first week of February; he was deeply touched that the mother had written him the night before delivery; she apparently had a difficult time (letter T27) and Vincent declared that her obvious courage helped him forget the last days when he was ill (letter 625). In the next week he painted *Blossoming Almond Tree* especially for his young nephew (letters 627 and W 20).

Initial respect for Dr. Peyron lessened as doubts about the usefulness of confinement increased. Months later Vincent recalled that his nightmares, under Peyron's well-meaning but ineffectual treatment, were somewhat aggravated (letter 640). Vincent felt that early concessions were not maintained and that the doctor did not distinguish his case from those of the demented inmates. In a letter to his mother he recalled that "my last attack [was] partly due to the influence which the illness of others had on me" (letter 639). The asylum also housed nuns and priests and Vincent felt that the background of the institution, even the cloisters around the chapel, were contributing to his "frightful ideas about religion such as never came into my head in the North." He was thinking that if subsequent attacks were "fit[s] of religious exaltation" then he should indeed consider moving away from the South (letter 607). The possibilities that were explored included boarding with the family of a sympathetic artist such as Camille Pissarro, Victor Vignon, or Auguste Jouve (letter 607), the amateur artist Dr. Paul Gachet at Auvers (letter 609), or staying at an asylum near Paris, something like the institution at Montevergues where they favored work therapy (letter 623). Dr. Peyron was still doubtful about the wisdom of a change and Vincent detected that the doctor spoke "vaguely so as to escape responsibility" (letter 631). They finally elected to follow the suggestion of Pissarro by seeking boardinghouse accommodation in Auvers-sur-Oise under the observation of Dr. Gachet.

Theo and Dr. Peyron worried about Vincent's safety during the proposed trip and thought that he should have an escort but this was rejected by the patient. Their concern was based on a misinterpretation of the effects of visits to Arles. It is worth mentioning that the attacks which followed those sojourns had nothing to do with motion sickness or changes of society and scenery as Hulsker[4] has speculated. Travel per se was not deleterious; the long overnight journey to Paris was completed without mishap.

Vincent kept up the pressure for release; Theo wrote the letter; Dr. Peyron acquiesced. The artist related to his mother a month later that he and Dr. Peyron "had words over it but we separated on good terms" (letter 639). Yet when permission was granted the artist was preoccupied with a last round of paintings and experienced great difficulty in organizing his affairs or packing his gear (letter 633). Dr. Peyron wrote "guérison" [cured] in the last column of the register at St. Rémy [this must be one of the best examples of wishful thinking in medical history – it was less than three weeks since the resolution of Vincent's last and longest crisis]. Vincent departed St.

Rémy on Friday, May 16, 1890, and sent a telegram to Theo from Tarascon before boarding the overnight train. He arrived in Paris mid-morning the next day and Johanna van Gogh-Bonger met for the first time "a sturdy, broad-shouldered man with a healthy color, a smile on his face, and a very resolute appearance." Her second observation was that Vincent looked much stronger than Theo.[1]

Paris

The next day Vincent was reunited with his own works on Theo's walls, also unframed canvases and even, as he mentioned later (letter 640), piles under the bed. The brothers also visited Père Tanguy's shop where Theo had rented a room to store some of Vincent's paintings. Vincent renewed his friendship with Johanna's brother Andries, but it seems unlikely that he contacted any artists at this time. Johanna van Gogh-Bonger[1] remembered that Vincent was insistent on eating olives every day; this habit apparently started after Mr. and Mrs. Ginoux had sent him boxes of olives while he was in St. Rémy (letter 622a). Vincent asked Theo for a letter of introduction to Dr. Gachet and left for Auvers-sur-Oise after just three days; he explained later that, "all the noise there [in Paris] was not for me" (letter 635). Auvers was Vincent's last domicile; Figure 2.9 is a summary.

Auvers-sur-Oise

Vincent was impressed with the little country town that had been home to Charles-Francois Daubigny and motif to Jean-Baptiste Corot. Honoré Daumier had lived a little north at Valmondois. This trio of artistic giants, who had been so influential on Vincent, died in the late 1870's within four years of each other [their graves are clustered in Père Lachaise Cemetery in Paris]. Jules Dupré had died in nearby l'Isle-Adam in 1889 and Camille Pissarro had spent some years downstream at Pontoise. Paul Cézanne enjoyed two periods in Auvers and Paul Gauguin had visited some years before Vincent, mostly to work with Pissarro. Armand Guillaumin still frequented the area. Dr. Gachet made his home there in 1872. Marc Edo Tralbaut, the van Gogh commentator who is frequently referenced in this book, was born in Auvers-sur-Oise in 1902.

Paul Gachet attributed his early nurturing of artistic tastes to the fine art museum in Lille and to a lasting friendship with Armand Gautier, a successful salon painter who migrated to Paris with Gachet in the 1840's. While at Montpellier, writing and defending his M.D. thesis,[22] Gachet became famil-

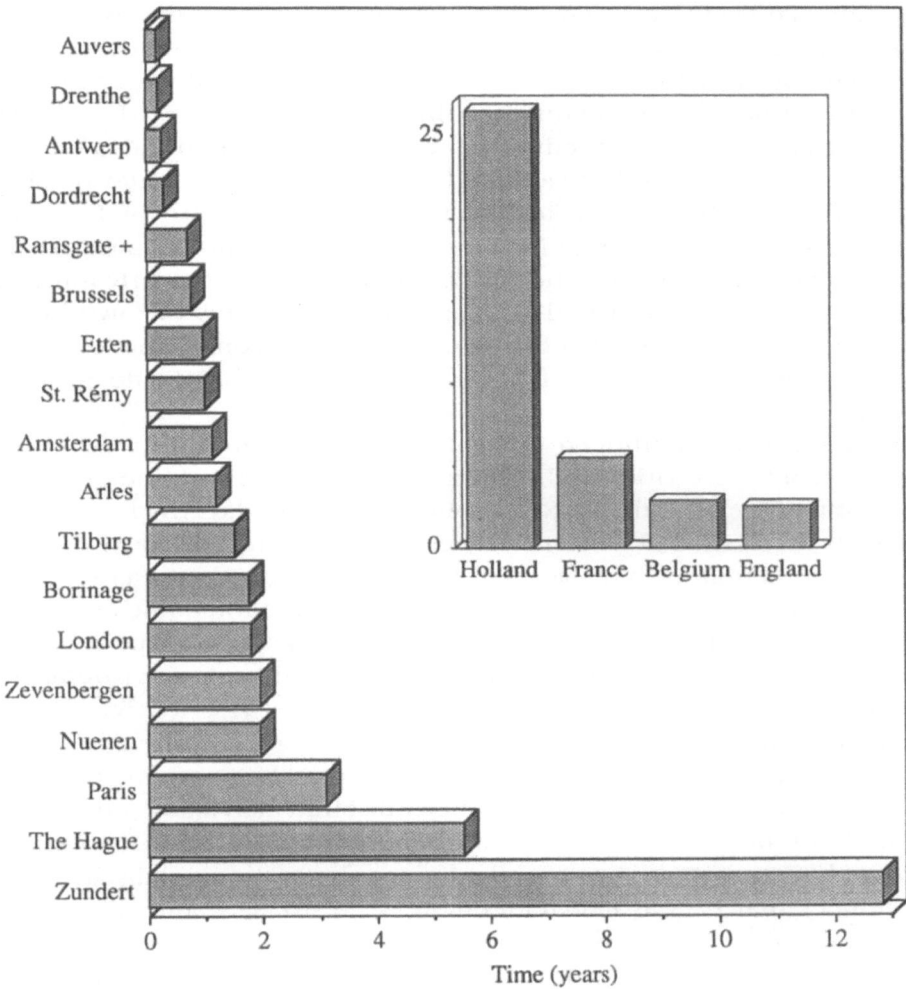

Figure 2.9. The domiciles of Vincent van Gogh. The locations have been rank ordered according to duration. Multiple residencies at the same venue have been summed. "Ramsgate +" stands for: Ramsgate, Isleworth, Turnham Green and environs. *Insert:* Domiciles by country.

iar with the Bruyas collection, the same display which had captivated both Gauguin and Vincent in 1888. From 1873 he practiced etching in his spare time and in his attic studio three friends, Pissarro, Guillaumin, and Cézanne combined with him to pull many prints. The same press was used for Vincent's only etching, *Man with a Pipe,* which was executed on the first afternoon that van Gogh visited Gachet for lunch (25 May 1890).

Dr. Gachet prescribed work therapy for Vincent; this was in the best tradition of one of Gachet's mentors, J. P. Falret. In a letter to Theo, Vincent repeated the doctor's advice: "I must work boldly on, and not think at all of what went wrong with me" (letter 635) and a little later to the Ginoux family, "The doctor here says that I ought to throw myself into my work with all my strength, and so distract my mind" (letter 640a). This was music to Vincent's ears. On the other hand he rejected Gachet's attempts at better nutrition: "the good soul [Gachet] takes the trouble to have four- or five-course dinners, which is as dreadful for him as for me – for he certainly hasn't a strong digestion [either] (letter 638). There is one particularly intriguing comment, "[Dr. Gachet] said ... that if the melancholy or anything else became too much for me to bear, he could easily do something to lessen its intensity" (letter 637). We would dearly like to know what the doctor had in mind!

I have already mentioned Vincent's inability or unwillingness to heed advice on better nutrition and cutting back on alcohol and smoking. There was not much else available to any physician of the day which could have reversed the course of Vincent's illness. It is therefore surprising that so many commentators have readily embraced Vincent's criticisms of Dr. Gachet without properly evaluating Vincent's judgment. Some of the more critical remarks by van Gogh included: "[Dr. Gachet's] experience as a doctor must keep him balanced enough to combat the nervous trouble from which he certainly seems to me to be suffering at least as seriously as I" (letter 635); "He certainly seems to me as ill and distraught as you or [I]" (letter 638); and the most quoted of all, "he is sicker than I am, I think, or shall we say just as much, so that's that. Now when one blind man leads another blind man, don't they both fall into the ditch" (letter 648). As much as one may admire Vincent's intelligence, and inherent observational skills, it is well to remember that he was intermittently very ill, stressed, and subsequently depressed enough to commit suicide. I think that it is foolhardy to rationalize his dismissal of Dr. Gachet's ability and advice, and for that matter some of the suggestions of Drs. Peyron and Rey, as patient insight and better judgment than his doctors. It was rather the culmination of Vincent's growing despair.

Vincent resided at a little café with boarders run by Arthur Gustave Ravoux. The building, which has been slightly altered over the years, faces the town hall of Auvers, a small town then as now, where one finds very attractive countryside just thirty kilometers from the heart of Paris. Gachet steered van Gogh to a better establishment, Café-Auberge Saint-Aubin, closer to the doctor and quieter, but Vincent deemed the 6 francs per day too much and settled for the smaller house that was only 3.5 francs. Whether this particular turned Ravoux against Dr. Gachet is not known. The Ravoux family had settled in Auvers only a short time before Vincent arrived, but it is clear that they soon developed an animosity toward Dr. Gachet, and their subsequent stories about Vincent always seemed at odds with those of Gachet's son, Paul Louis Gachet. Ravoux lost much credibility by insisting that Vincent hardly saw Gachet and never ate at his house, which is absurd in light of Vincent's documentation and the fact that Vincent, Theo, Johanna, and young Vincent were all entertained at the Gachet domicile one Sunday afternoon.

It was probably the last happy time together for Vincent and Theo as they all enjoyed the day at the house of Dr. Gachet. Vincent introduced his little namesake to the animals and birds of the Gachet menagerie. Subsequent criticism of Dr. Gachet in popular articles even extended to one of his exotic pets, a peacock who was described as scrawny, damaged and unkempt. In fact the doctor had provided a safe haven for a discard from a circus.

Van Gogh's productivity in Auvers was immense, something of the order of a canvas a day (see Figure 2.10). He painted Gachet's daughter Marguerite in the garden and also at the piano. According to Gachet's son[23] the latter composition was executed shortly after his sister's twenty-first birthday (June 21) and was inspired by a Dr. Gachet etching depicting his late wife at the piano. Another study of the garden was described as having "an aloe with marigolds and cypresses" (letter 638). Actually it was a thuja tree, cypresses do not grow as far north as Auvers, but Vincent rendered it with the same flame-like swirls that had become his signature in the South. *The Church at Auvers,* and a multitude of village scenes and landscapes were executed with the same aplomb. His command of the brush was everywhere, and even within four days of a most disturbing visit to Paris, Vincent finished three pictures including *Daubigny's Garden* (letter 649).

The young baby was sick and Vincent declared that he would like to visit but felt "more powerless than you [Theo] in the present state of anxiety" (letter 646). Six days later (July 6) he caught an early train to Paris and spent the day with Theo, Johanna, young Vincent, Toulouse-Lautrec (who stayed for lunch), the writer and art critic Albert Aurier, Andries and Annie Bonger and possibly others.[1] The significant events, especially those concerning the future plans of Theo, have not yet been properly disclosed. There was obvious uncertainty about Theo's position at Boussod & Valadon. This would affect Vincent's future support by his brother if an outlay for an independent gallery was anticipated. Perhaps Johanna and Theo were having marital difficulties or Johanna and her sister-in-law, living in the same apartment building, had quarreled? We now know that Theo's health was about to take a turn for the worse. In any event Vincent wrote that "we are all rather distressed, and a little overwrought besides, ... You rather surprise me by seeming to wish to force the situation" (letter 647). According to

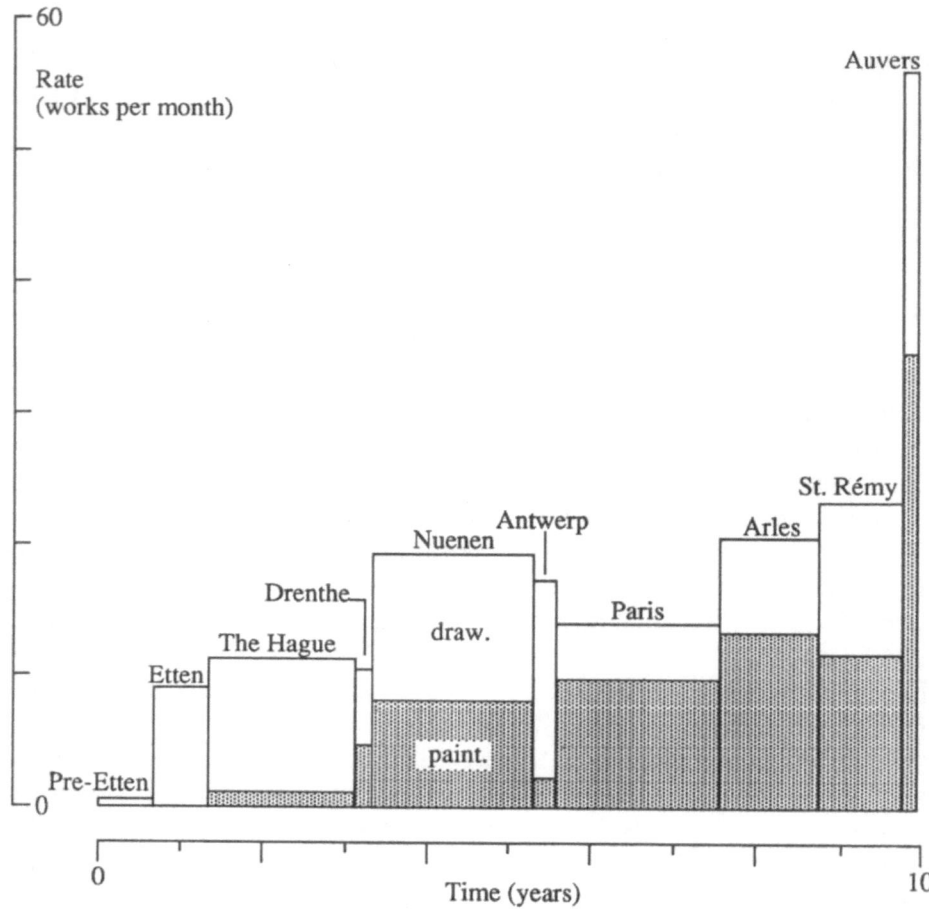

Figure 2.10. Art work: duration, intensity and production at ten locations. Each bar represents the *time* (width) in residence (for example, 1.02 years at St. Rémy) and the *rate* (height) of paintings (stippled panel) or drawings (clear) completed during that period. The area of each panel is thus proportional to the *production* (for example, an equal number of drawings and paintings were produced at St. Rémy). The periods were determined by the author. The item numbers were gleaned from Hulsker, *The Complete van Gogh, Paintings, Drawings, Sketches*, according to the following operational definitions. (1) Works on paper in chalk, pencil, charcoal, watercolor, gouache, or combinations thereof, were grouped under drawings. (2) Sketches in the text of Vincent's letters, or sent with letters, were not counted. (3) The paintings include both studies and major works. Trends in intensity and type of art work are clear. However, there are inherent restrictions, centered on questions of survival and attribution, which affect the absolute values.

Hulsker,[4] the published version of letter 647 is incomplete and his translation of the last sentence is "You rather surprise me by seeming to wish to force the situation [with me] while *there are disagreements between you* [Theo and Johanna]."[4] But Vincent received assurances and wrote that "Jo's [next] letter was really like a gospel [glad tidings] to me, a deliverance from the agony which had been caused by the hours I had shared with you ... [but] it was no slight thing when we all felt our daily bread was in danger ... our means of subsistence were fragile." He still felt very sad and "continued to feel the storm which threatens you weighing on me too." He feared "being a burden to [them]," and acknowledged that "Jo's letter proves to me clearly that you understand that for my part I am as much in toil and trouble as you are" (letter 649).

Back in Auvers, Vincent had an altercation with Dr. Gachet over a Guillaumin painting of a reclining nude. Van Gogh admired the picture and was disturbed about the doctor's failure to have it properly framed. According to Gachet's son[24] this led to an intense argument and the suggestion that Dr. Gachet had to control the artist with a strong directive (shades of Gauguin and the razor story). In any event, van Gogh was back to normal the next day.[24]

On July 11, Vincent wrote his mother that he was happy that she would be seeing the new grandchild shortly (letter 650). On July 14, Theo wrote Vincent that they were glad he was feeling "less dispirited on account of the unsettled business questions ... The danger is really not as serious as you thought" (letter T 41). Theo's family went to Leyden, the new location for mother van Gogh and sister Wil, the next day. Much has been written about Vincent's displeasure with Theo's household planning to spend their summer vacation in Holland instead of with him. And yet it is difficult to interpret because he continued to paint, e.g. *The Town Hall at Auvers on Bastille Day,* and on July 23 he was still ordering paints and reviewing pictures (letter 651). Vincent's productivity was impressive and he covered a broad range of subjects. The motifs for most of van Gogh's paintings in Auvers-sur-Oise have been identified and photographed by Mothe.[25]

Theo returned to Paris somewhat earlier than planned with the intention of rejoining his family in Holland sometime later. On July 20, he wrote his wife that he hoped that Vincent was "not getting melancholy or that a new attack is threatening again," and on July 25, that he had "a letter from Vincent which seems quite incomprehensible."[1] It was incomprehensible to Theo because he had decided to remain with his employers, according to an unpublished letter to his mother,[4] but he had neglected to tell Vincent. Apparently the assurance of letter T41 had not dispelled Vincent's fears on this score.

The short script found on Vincent, and published in *The Complete Letters* as letter 652, may have been a draft of letter 651.[4] If this is so then the vision of impending doom, "Well, my own work, I am risking my life for it ..." (letter 652) was excluded from the final version (letter 651). (These aspects are discussed in more detail in chapter 10.) Accordingly, the most significant parts of the putative last letter may be, first, "I hope that you will have found those worthy gentlemen [Boussod & Valadon] well disposed toward you" [note that Vincent was unaware of Theo's decision to stay with B.& V.] and, second, that he ordered paints for his new colleague Anton Hirschig and himself albeit, "I have reduced my own order to the barest minimum." The close, "Good-bye now, and good luck in business, etc., remember me to Jo and handshakes in thought" certainly puts a happy face on a "final communication," if indeed that was intended.

Vincent died about 1:30 a.m. on July 29, 1890 as a consequence of a gunshot wound about thirty-three hours earlier. That much is clear – everything in between is fraught with conflicting stories and subject to various interpretations. Many of the secondary players in the drama were seduced into offering stories only after Vincent van Gogh and his art work achieved recognition. For several, the passage of time improved the significance of their own parts. (The animosity of the Ravoux family towards the Gachets has already been mentioned.)

Dr. Mazery, a local physician, was called by Gustave Ravoux after he discovered that Vincent had shot himself in the afternoon or early evening of July 27, and had returned to his room bleeding. (It is remarkable that Ravoux should go so far as to avoid calling Dr. Gachet.) Vincent asked for Dr. Gachet. The two doctors conferred and elected not to remove the bullet, but dressed the wound and made the patient as comfortable as possible. Johanna van Gogh-Bonger[1] is clear on the following; the next morning Dr. Gachet sent Anton Hirschig to Paris with a message for Theo, who jumped on a train and reached Vincent that same morning, July 28.

The intent of suicide was seemingly confirmed. When Dr. Gachet told Vincent he hoped to save his life the victim supposedly said "then it has to be done all over again." Vincent died in the early morning of July 29 and the death certificate (municipal archives of Auvers-sur-Oise) was signed by the mayor (Alexandre Caffin), Gustave Ravoux, and Theo van Gogh (the document is reproduced in Tralbaut;[3] neither doctor signed). Theo had a notice of the funeral, scheduled for 2:30 p.m. on July 30, 1890, printed in Pontoise and distributed in Paris.[16]

The Gachets and others assisted Theo in arranging the coffin, flowers, and van Gogh paintings about the small reception room in the café. Given the short notice it is remarkable how many friends arrived for the funeral; three years later Émile Bernard painted the scene from memory[3] and included himself, Theo van Gogh, Charles Laval, Andries Bonger, Lucien

Pissarro, Auguste Lauzet, Père Tanguy, Gustave Ravoux, Dr. Paul Gachet and others. Worried again by adversity, even beyond death, Vincent's body was refused transit on the parish hearse because he was a suicide, and the remains travelled up the hill to the small cemetery on a cart borrowed across the river at Méry-sur-Oise.

In his account to Albert Aurier, Émile Bernard remembered how they lowered Vincent into the grave and that Dr. Gachet attempted some words of farewell which were halting because he too was weeping. In a letter to his mother Theo sorrowed that, "It is a grief that will last and which I certainly shall never forget as long as I live [he died within six months of his brother, after experiencing bouts of illness with symptoms resembling the same underlying disease]; the only thing one might say is that he [Vincent] himself has the rest he was longing for." Even then the saga was unfinished.

The initial concession on Vincent's burial place was for fifteen years and nonrenewable. The van Gogh family was advised, and Johanna van Gogh-Bonger duly made the official application for exhumation and reburial in a larger plot. The approval was issued June 9, 1905 in the name of Madame Cohen-Gosschalk-Bonger (she had married Johan Cohen Gosschalk in 1901) and the process was carried out four days later. Paul L. Gachet assisted with arrangements. In 1914, the remains of Theo were transferred from Utrecht, Holland to lie beside his brother Vincent. Dr. Gachet's son provided an endowment for the perpetual upkeep of the graves, which remain covered with ivy as requested by Johanna van Gogh-Bonger. The headstones are simple; on the left, "ici repose (here reposes) Vincent van Gogh (1853–1890)" and on the right, "ici repose Theodore (sic, should read Theodorus) van Gogh (1857–1891)."

References

1. van Gogh-Bonger J. 1978. Memoir of Vincent van Gogh. vol I, XV–LIII, in: *The Complete Letters of Vincent van Gogh.* 2nd ed. Boston: New York Graphic Society.
2. Sanders P. 1981. Genealogie van Gogh. *Genealogisch Tijdschrift voor Midden-en West-Brabant* 5(4): 233–251.
3. Tralbaut ME. 1981. *Vincent van Gogh.* New York: The Alpine Fine Arts Collection Ltd.
4. Hulsker J. 1990. *Vincent and Theo van Gogh: A Dual Biography.* Ann Arbor: Fuller Publications.
5. van den Eerenbeemt HFJM. 1971. Van Gogh in Tilburg. *Brabantia,* November 1971, quoted in Hulsker (ref. 4).
6. du Quesne-van Gogh E. 1913. *Personal Recollections of Vincent van Gogh.* (translated by KS Dreier) Boston & New York: Houghton Mifflin Company.
7. Wood GB & Bache F. 1865. *The Dispensatory of the United States of America.* Philadelphia: JB Lippincott & Co.

8. van der Wolk J. 1987. *The Seven Sketchbooks of Vincent van Gogh: a Facsimile Edition.* (translated by C Swan) New York: Harry Abrams Inc.

9. Cachin F & Welsh-Ovcharov B. 1988. *Van Gogh à Paris.* Paris: Editions de la Réunion des Musées Nationaux.

10. Sheon A. 1978. *Monticelli: his Contemporaries, his Influence.* Pittsburgh: Museum of Art, Carnegie Institute. (see especially Chapter 7, "Monticelli and van Gogh," pp 81–91.)

11. Stein SA. 1986. *Van Gogh: a Retrospective.* p 87. New York: Park Lane.

12. Guigou P & Lauzet A. 1890. *Adolphe Monticelli.* Paris: Boussod & Valadon.

13. Pickvance R. 1984. *Van Gogh in Arles.* New York: Harry N. Abrams Inc.

14. Letters to Émile Bernard (B1 – B22). 1978. vol III, pp 473–527, in: *The Complete Letters of Vincent van Gogh.* 2nd ed. Boston: New York Graphic Society.

15. Barr AH Jr. 1936. *Vincent van Gogh.* pp 44–46. 3rd ed. New York: The Museum of Modern Art.

16. Pickvance R. 1986. *Van Gogh in Saint-Rémy and Auvers.* New York: Harry N. Abrams Inc.

17. van Uitert E, van Tilborgh L & van Heugten S. 1990. *Paintings: Vincent van Gogh.* Milan: Arnoldo Mondadori Arte srl.

18. van der Wolk J, Pickvance R & Pey EBF. 1990. *Drawings: Vincent van Gogh.* Milan: Arnoldo Mondadori Arte srl.

19. Gauguin P. 1949. *Paul Gauguin's Intimate Journals.* (translated by VW Brooks) New York: Liveright Publishing Corporation.

20. Cocquiot G. 1923. *Vincent van Gogh.* Paris: Ollendorf. This is also quoted as item 590a in *The Complete Letters of Vincent van Gogh.*

21. Arnold WN. 1988. Vincent van Gogh and the thujone connection. *Journal of the American Medical Association* **260**: 3042–3044.

22. Gachet P-F. 1858. *Étude sur la Mélancolie.* Montpellier: Éditeur du Montpellier Médical.

23. Gachet P[L]. 1953. *Van Gogh à Auvers. Histoire d'un Tableau.* Paris: Les Beaux Arts, Edition d'Etudes et de Documents.

24. Gachet P[L]. 1956. *Deux Amis des Impressionnistes: le Docteur Gachet et Murer.* Paris: Éditions des Musées Nationaux.

25. Mothe A. 1987. *Vincent van Gogh à Auvers-sur-Oise.* Paris: Éditions du Valhermeil.

Chapter 3
Illnesses

Figure 3.1. Vincent van Gogh: *The Sower*, 1888, Arles, oil on burlap mounted on canvas, 73 × 92.5 cm (28.7 × 36.4″), Foundation E.G. Bührle Collection, Zürich

*I did say [to Dr. Rey] that I myself should
always regret not being a doctor.*
Vincent to Theo, letter 571, from Arles,
January 17, 1889.

Vincent gave clear descriptions of his maladies and his bouts of ill health. The depth to which he analyzed both the comments of his doctors, and the daily items affecting his well being, was quite remarkable. Furthermore, the artist entered into open discussions with his brother Theo, with Émile Bernard, and to some extent with his mother and one sister. The letters are candid, there is a high degree of self-exposure. This chapter is primarily a précis of Vincent's own descriptions, which have been organized into a framework for later analyses. Some comparisons with twentieth century medical practice are made *en passant*. For each category of illness or disorder I have selected representative examples from the letters, which are identified by number, and then grouped them into tables of related items. The chapter closes with brief discussions on the illnesses of Theo and Wil.

Infectious diseases

We can assume that Vincent had his share of the usual coughs, colds, and earaches that beset most children although nothing was outstanding enough to be remembered by his sister Elisabeth,[1] or relayed by his mother to Johanna van Gogh-Bonger.[2] However, there is no satisfactory explanation for the unfinished concluding term of high school at age fifteen and, in the absence of a better understanding of this year and the next, there is the possibility of undocumented illness. It is somewhat paradoxical that Vincent's mother felt that he was the most robust of the young children.[2]

The first documented, debilitating infection was contracted at The Hague in mid-1882 (age 29) in the form of gonorrhea or, as Vincent called it, the clap (letter 206). The disease was diagnosed in a large hospital by experienced physicians, and there can be little argument about accuracy. According to Holmes[3] the usual incubation period in males is 2 to 7 days so that the earlier episodes of headache and fever in January that year (letter 172) and an intervening fortnight of malaise and fever (letter 200) were probably not related to the June hospitalization. Before the advent of antibiotic treatments, the symptoms of gonococcal urethritis persisted about eight weeks on the average.[3] Thus the length of his stay in the hospital – June 7 to July 1 – was reasonable, but the symptoms seemed to persist until about the fourth week of October (letter 238), which would indicate a protracted case. The resolution was natural and may have been aided by quinine for the fever and irrigations of the urethra with solutions of alum (according to Vincent) or zinc sulfate (hospital register). The infection itself, as well as the trauma due to the insertions of catheters (letters 208, R 10) to drain the bladder and to irrigate the urethra, resulted in scar tissue and a decrease in the caliber of the urethra, i.e. a stricture. Consequently, Vincent experienced problems with urination, which the physicians attempted to relieve by inserting bougies,* instruments of increasing girth, in order to dilate the urethra; this was a painful process (letter 209).

References to infectious diseases, related symptoms, and procedures

Gonorrhea: 206–209, 212–216, 218, 221, 233, 238, 569
Respiratory tract infection: 448, 601
Syphilis: Tralbaut p. 177
Urine retention or tract obstruction: 206, 208, 209, 214, 215, R10
Dysuria: 206, 214, R10
Catheter: 208, R10
Bougies: 209

In the past, gonorrhea infections were probably the most common cause of urethral stenosis, but the present treatment of such inflammations is unlikely to be associated with fibrosis so that strictures are seen less frequently today.[4] Dilation is still a primary treatment for urethral strictures; it is done gradually and may require several visits over weeks to dilate to 24F,* and then may require further treatments six months or a year later. Vincent's course was shorter, may have been incomplete or even caused more harm than good, and possibly allowed a recurrence of urinary tract problems in Antwerp (1885–86).

Syphilis and gonorrhea are often concomitant infections[3] and, although Vincent may have contracted both in 1882, the former may have escaped detection at the primary stage. Several commentators have assumed that Sien Hoornik, a former prostitute and Vincent's lover in The Hague, was a carrier of venereal diseases. (A significant proportion of women with gonorrhea never develop symptoms[3] and primary syphilis is often overlooked in

---

*   Bougies are metal probes with calibrated, olive-shaped tips which are used primarily for gauging the size of the urethral opening. Vincent mentioned bougies, but a more commonly used instrument for dilations of the urethra is a curved rod called a sound. In both cases the external diameters follow the so-called French scale wherein 1F is equivalent to 0.33 mm.[4]

women.[5]) Sien was pregnant when she met Vincent. Congenital syphilis can sometimes lead to abortion, stillbirth or neonatal death.[5] However, because a latent infection may not become expressed till adulthood,[5] her apparently healthy child does not rule out the possibility that Sien harbored syphilis. Notwithstanding the numerous claims in the literature that van Gogh had syphilis the modicum of documentation comes from Tralbaut's interview[6] with the descendents of Dr. Hubertus Cavenaille, who insisted that their grandfather had treated the artist specifically for syphilis in Antwerp in 1885–86. In contradistinction, there is no mention of syphilis in Vincent's letters, and Cavenaille's name appears in Vincent's sketch book[7] beside notes on alum irrigations and sitz baths, which are reminiscent of his previous treatments in The Hague for gonorrhea.

Vincent was also subject to infectious diseases of the respiratory tract which were not extraordinary. In Antwerp, in February 1886, he "began to cough continually ... and to expectorate a grayish phlegm" (letter 448). The cough suggests lower respiratory tract involvement and a productive cough usually means a bacterial infection (or a viral infection with a secondary bacterial infection) involving pus. Three years later in St. Rémy he was unable to eat for four days "because of a swollen throat" (letter 601). A viral or bacterial throat infection was indicated.

Infections can precipitate other problems upon an appropriate underlying condition and in this context the "complications" during the third week of the gonorrhea attack (letter 208), including urine retention, a vague mental state "more or less like a dream," and complaints of nervousness, are worth noting for later consideration.

Malnutrition

With regard to nutrition, the artist's habits were wretched, his general status was poor, and even his expectations were limited. The problem was partly fiscal because he budgeted poorly and elected to spend his money on artist's materials rather than ordinary foodstuffs. At times he was so preoccupied with painting in the field that he neglected to pack something to eat. Household chores such as cooking were never attractive. His food intake was notably decreased when he lived alone, and he sometimes found it easier to imbibe calories from a bottle. On occasion, dental caries and toothache made chewing painful, and bouts of illness provoked anorexia, or aroused gastrointestinal complaints. All resulted in subsequent periods of malnutrition. The letters are filled with Vincent's comments about needing to get some nutritious food into his body, concern about simply not eating enough, and worry about finding the right food in restaurants.

References relating to malnutrition

Fasting, & irregular intake: 304, 308, 310, 440, 442, 449, 454, 457, 509, 546, 571, 581

Poor nutrition: 306, 308, 310, 442, 454, 480, 581a, 590b

Weight loss: 444

Decreased intake secondary to:
indigestion 316, 469, 480
alcohol 590b
throat infection 449, 601
dental caries & toothache 173, 255, 448, 449, 450, 455

Anorexia: 215, 307, 469

Feelings of faintness, malaise, and nervousness in Antwerp, December 1885, were directly correlated with fasting, by the artist himself (letters 440, 442). It should be noted that for part of January–February 1889, after his first major crisis in Arles, Vincent sustained a fast and disdained Dr. Rey's directions to eat more (letter 571); this probably contributed to his second crisis.

An unsettling theme in Vincent's correspondence is his philosophy that sickness, or something less than full health, is compatible with, or even encourages, creative work. In this vein, from Antwerp in 1886, he wrote, "one must not think that people whose health is impaired, wholly or partly, are no good for painting. ... *one need not be perfectly healthy* ... [in any event] nervous people are more sensitive and refined" (letter 449, my italics here and elsewhere in the chapter); and from St. Rémy, end of 1889, "it isn't necessary to be a Hercules to stick assiduously to the easel" (letter 604). In Arles, January 1889, he wondered if sickness might even be necessary for highly creative work because, "if I recover, I must begin again, and I shall not again reach the *heights to which sickness partially led me*" (letter 570). Finally, and perhaps with the most devastating consequences for his under-lying disease, he shunned doctors' orders about eating properly, "*As for eating a lot,* [at present] I do – but if I were my [own] doctor, *I'd forbid it. I don't see any advantage* for myself in enormous physical strength, because it would be more logical for me to get absorbed in the thought of doing good work and wishing to be an artist and nothing but that" (letter 607).

Alcohol abuse

Vincent was raised by a mother and father who were moderate in every way; their attitude to alcoholic beverages may have bordered on the puritanical.

In any event they were deemed by the artist to have, "prejudices and old-fashioned ideas which neither you [Theo] nor I can share anymore. ... they bring up a story of a great uncle who was infected with French ideas and *took to drink,* and so they *insinuate that I shall do the same.* Quelle misère!" (letter 159, November 1881). Evidence that Vincent van Gogh was a drinker came from friends, his doctor, a group of disturbed neighbors, Theo, and the artist himself.

The earliest indication of Vincent's affinity for alcoholic beverages was related by Anton Kerssemakers from 1884 (the Nuenen period). This amateur artist and friend remembered that Vincent preferred simple food-stuffs, cheese and unbuttered bread, and declined to eat ham or put sugar in his coffee on the grounds that it would be "coddling [him] self too much." [but] "On the other hand, he [Vincent] liked to have some *brandy in his flask* on his rambles [painting excursions], and he would not have liked to do without it ... this was the only luxury" (letter 435c). Van Gogh started drinking excessively after his arrival in Paris in 1886. Most of the letters that bear on his life style in Paris were written later but they make the point over and over again.

References to alcohol

Affinity:
    Nuenen 435c
    Paris 599
    Arles 474, 599, W15
Overindulgence:
    Paris 480, 481, 544, 544a, 611
    Arles 507, 514, 585, 590b, A16, Tralbaut pp. 269–270
Alcohol to the exclusion of food: 581, 590b
Projections:
    Monticelli 478, 481, 507, 550
    Roulin 550
    Bismarck 514
    "Marseille Artist" 588
Absinthe:
    Others 470, 478, 588
    Vincent A16
Advice and philosophy:
    Dr. Rey 581a
    Self 480, 534, 583b, 585, 595, 599

From Arles, Vincent wrote to his brother, "I am better than I was in Paris, and if my stomach has become terribly weak, it's a trouble I picked up there and most likely due to bad wine, which I *drank too much*" (letter 480) and "I was certainly going the right way for a stroke [breakdown] when I left Paris. I paid for it nicely afterward! *When I stopped drinking*, when I stopped smoking so much, when I began to think again instead of trying not to think – Good Lord, the depression and prostration of it" (letter 481). He recalled to Theo, "When I left you at the station to go South, [I was] very miserable, almost an invalid and *almost a drunkard*" (letter 544) and to Gauguin and Schuffenecker, "when I left Paris, [I was] seriously sick at heart and in body, and *nearly an alcoholic*" (letter 544a). From St. Rémy, Vincent compared himself with Isaäcson who "stayed in Paris a long time [but] I suppose he is *wiser than I, and has not been drinking*" (letter 611).

It is clear that he resumed in Arles. There was a justification, "the only thing to bring ease and distraction, in my case and other people's too, is to stun oneself with *a lot of drinking* or heavy smoking" (letter 507). And then there was the worry, "I myself have become haggard of late ... As for *drinking too much* ... if it is bad, I can't tell" (letter 514, July 29, 1888). (This apparently answered a query from Theo about his current drinking habits!) Here, and elsewhere, Vincent continues to analyze his problem, "you do understand that *if alcohol has undoubtedly been one of the great causes of my madness* then it came on very slowly and will go away slowly too, assuming it does go of course ... [but be aware of] the frightful superstition of some people on the subject of alcohol so that they [actually] prevail upon themselves never to drink" (letter 585, April 21, 1889).

After his second medical crisis in Arles some neighbors petitioned[8] the mayor to lock him up, maintaining that "he indulges in *excessive drinking*, after which he finds himself in such a state of excitement that he *no longer knows either what he does or what he says*." Tralbaut quotes the municipal librarian at Arles, "He [Vincent] never made a scandal, except when he had been *drinking, which happened often*."[9] There were also indications that alcohol was consumed to the exclusion of other foods, "[Dr.] Rey says that instead of eating enough and at regular times, I *kept myself going on coffee and alcohol. I admit that*" (letter 581, March 24, 1889). Dr. Rey also described how Vincent neglected cooking and noted that after a hard day in the field, "he consumed the hardly inviting food [half-cooked chick peas], unless he preferred to *drink spirits* to relieve his stomach" (letter 590b).

Vincent does not discuss the particular spirits he drank in Arles although references to other people drinking absinthe in the South are made in letters 470 and 588 and we know that the region had four times the national per capita rate of absinthe consumption.[10] Signac gives us a clue, "Though he [Vincent] ate hardly anything, *what he drank was always too much*. Returning after spending the whole day in the blazing sun ... he would take his seat on the terrace of a café. And the *absinthes and brandies* would follow each other in *quick succession*" (letter item A16). Absinthe and its toxic components in addition to ethanol are discussed in the next chapter.

There are several examples where Vincent discusses the hard drinking of others including Monticelli (letters 478, 481, 550), Bismarck (letter 514), Roulin (letter 550) and a "Marseille artist" who committed suicide but, Vincent protests, *"not ... as the result of absinthe"* (letter 588). In this last group of letters Van Gogh attempted to rationalize drinking habits generally, projected his own problem onto others, and tried to excuse it. His sensitivity to brandy, "one small glass makes me tipsy" (letter 474) in the early months in Arles does not indicate that he now abstained but rather shows that he was still tippling while offering sanctimonious advice to Theo.

He was advised not to drink by Dr. Rey (letter 581a) and admitted that his survival would depend on sobriety (letters 583b & 585). Finally, we have two personal declarations of the problem, "I live soberly [now] because I have a chance to, I *drank in the past* because I did not quite know how to do otherwise. Anyway I don't care in the least!!!" (letter 599, July 5, 1889), and, to his sister Wil, "The physician ... did not consider me a lunatic, but that the crises I have are of an epileptic nature. Consequently [if we buy the epilepsy hypothesis] *alcohol* is also not the cause, though it must be understood that it *does me no good either*" (letter W15, October 1889). His philosophy was interesting and sometimes even humorous, "Very deliberate sobriety ... leads nevertheless to ... thoughts, *if you have any,* [which] move more readily. In short it is a difference [between] painting in gray or in colors. ... I have a feeling [now] rather like I had when I was younger when I was very sober, *too sober they used to say then,* I think" (letter 599).

Neurological problems

Van Gogh experienced many forms of neurological disturbance. The most devastating, by his own admission, were the hallucinations, both auditory and visual. He was probably also hallucinating before or during the ear-cutting episode (see chapter 9) of the first crisis in Arles, but he had no recall of the event; total amnesia. Months later he summarized to sister Wil, "I have had in all four great crises during which I *didn't in the least know what I said, what I wanted, and what I did.* [Also] I had previously had *three fainting fits* without any plausible reason, and without retaining the slightest remembrance of what I felt" (letter W11, April 10, 1889). In Arles, he experienced further bouts but, "the unbearable *hallucinations* have ceased, and are now getting reduced to a simple *nightmare,* in consequence of taking bromide of potassium I think" (letter 574). He may have been hallucinating when he shot himself in Auvers-sur-Oise (see chapter 10).

When Vincent voluntarily entered the asylum at St. Rémy the attending physician recorded that he was suffering from *acute mania with hallucinations of sight and sound*. This was based upon medical reports from the Arles hospital and the initial patient interview. In a letter to Theo shortly thereafter, Vincent compared his situation, "I gather from others that during their attacks they have also heard strange sounds and voices as I did, and that in their eyes too things seemed to be changing. ... when it *comes on you unawares*, [it] cannot but frighten you beyond measure. ... For the anguish and suffering are no joke once you are caught by an attack. ... in my case it was *my sight as well as my hearing*, ... the shock was such that it *sickened me even to move*, and nothing would have pleased me better than never to have woken up again" (letter 592). The attacks were precipitous, "I had been working perfectly calmly on some canvases ... suddenly, without any reason, *the aberration seized me*" (letter 620). For the most part, Vincent was incapacitated by his attacks in Arles and St. Rémy, but on rare occasions he worked on through them and later found that, "some of my pictures certainly show traces of having been painted by a sick man, and I assure you that I don't do this on purpose" (letter W 16, November 1889).

References to neurological problems

Hallucinations: 574, 576, 592 607, 620, St Rémy Asylum register
Delirium: 623a, W17
Nightmares: 574, 602a, 613, 640, W4
Fainting fits: W11
Epilepsy: 589, 591, 592, W15
Stupor: 489, 628
Absent mindedness: 586, 601
Amnesia: W11
Loss of consciousness: 610
Headache: 173, 234, 308
Back pain: 305
Fits with religiosity: 607, W17
Dizziness: 308
Impotence: 506, 590, B14
Crises: 610, 620, 623a, 624, 628, 629, 634a, T27, T32, W11
Aphasia:
 Roulin to Theo, 28 December 1888
 Dr. Peyron to Theo, 29 January 1890
Neurosis: 481, 492, 513, 556, 576, 582, 585, 601, 635, 650, W4, W22

The nature of Vincent's seizures remains an open question. It is clear that both the patient and his doctors felt that he had some sort of epilepsy: "the doctor [Peyron] here is inclined to consider what I have had [to be] some sort of epileptic attack" (letter 591). He even attempted to rationalize the ear-cutting episode, "Most epileptics bite their tongue and injure themselves. [Dr.] Rey [Arles Hospital] told me that he had seen a case where someone had mutilated his own ear as I did" (letter 592). To the best of my knowledge there is no evidence in the literature to justify this comparison. When tongue-biting occurs it is accidental to the seizure; I suppose that ear lobe mutilation could occur by accident only if the patient had a straight razor (see chapter 9), or a pair of scissors in his hand at the onset of the attack.

Frank seizures were not described for Vincent. The "three fainting fits" he mentioned to his sister (letter W11), albeit difficult to construe as seizures, may have influenced Drs. Rey and Peyron in that direction. (Primary distinctions are now made among the types of seizures; for example *petit mal seizure, grand mal seizure, complex partial seizures,* and *temporal lobe epilepsy,* and see chapter 6.) As discussed earlier, and in chapter 6, it seems plausible that the physicians were unduly influenced by the patient's claim of a family history of epilepsy. To the best of my knowledge the extent of this so-called family history has never been properly documented.

Vincent does refer to the concept that he and Theo, and perhaps even Wil, had an inherited disease. Thus he wrote to Theo, "My poor boy, *our neurosis,* etc., comes, it's true, from our way of living, which is purely the artist's life, but it is also a *fatal inheritance,* since in civilization the weakness increases from generation to generation. .... *a neurosis which already has its roots in the past"* (letter 481). Vincent compared himself with Dr. Gachet in his nervous condition (letter 635) and later included Theo in the unhappy trio, "He certainly seems to me as ill and distraught as *you or me"* (letter 638). Earlier, Vincent included sister Wil in the family propensity for a *"nervous temperament"* (W 4), and expressed concern to Theo that Wil might need to be provided for in the future (letter 589).

In the previous chapter I documented the episodic nature of Vincent's ailment during his last two years and I summarized diagrammatically (see Figure 2.8) the six major crises in Arles and St. Rémy. It is worth repeating here that the episodes of acute mental derangement and the attending disability were separated by intervals of lucidity and creativity. Toward the end of his last and longest crisis, starting at the end of February 1890, Vincent gave clear indication of confusion even into April, "Today I wanted to read the letters which had come for me [and were held by the administration during the crisis], but I was *not clear-headed enough yet to understand them"* (letter 628). He obviously put this letter aside and then took it up again in a lengthy postscript because, "my *head is so bad,* without pain it is true, but altogether *stupefied.*" He described the onset, "My work was going well ... And the next day, down like a brute. Difficult to understand" (letter 628). In the next letter (April 29, 1890) Vincent not only declared that he was feeling a bit better but presented evidence of proper cognition by congratulating Theo on his impending birthday (May 1), giving instructions for the maintenance of some canvases, ordering some more pigments, and philosophically reviewing the previous two months, "What am I to say ... Things didn't go well at all" (letter 629). Shortly thereafter he started negotiations for release. Many disease states will explain some of the problems mentioned above, but it is equally important to realize that several of these possibilities can be ruled out immediately on the basis that if they are untreated their course is essentially unremitting as opposed to intermittent.

Psychiatric problems

On many occasions Vincent spoke of excessive nervousness, melancholy, irritability and anxiety. Some examples from the letters are assembled in the table. I have included the connection with Hugo van der Goes here although the subject is discussed separately in the next section.

References to melancholia started as early as age 29 years (letter 212). Vincent was also prone to associate with others who complained of undue nervousness and depression, for example Émile Bernard (letter B7) and Christian Mourier Petersen (letter 490). He compared himself with Hugo van der Goes (letters 514, 556), Charles Méryon (letter 546), and a host of other artists who suffered from incapacitating depression. According to Vincent, Dr. Paul Gachet was also in the same boat (letter W22), although I have already expressed a different opinion about this evaluation.

References to psychiatric problems and responses

Melancholy: 212, B7, 440, 546, 583, 601, 637, 649

Excited or sluggish, noir-rouge, [black-red or depression-mania]: 302, 492, B21 [projection]

Concern for breakdown: 449

Insomnia: 200, 206, 233, 570

Irritability, nervousness, anxiety: 173, 208, 212, 233, 302, 344, 481, 556, 582, 590b, 635, 638, W11, W22

Hugo van der Goes connection: 514, 556

Pica for terpenes and related chemicals: 570, 590a, Hulsker p. 372 & p. 414, T23, T24

Thoughts of suicide: 588, 602a, 605

Religiosity: 605, 607

Comments suggesting an early demise: 309, 442, 449

Positivity of work therapy: 489, 513, 582, 595, 596, 640a

The symptoms ranged from an "undercurrent of vague sadness" (letter 583) to "moods of indescribable mental anguish" (letter 582) and "horrible fits of anxiety" (letter W11). The artist always felt less prone to melancholia while he was working, "I do not feel faint as long as I am painting, but in the long run those intervals are sometimes rather too melancholy" (letter 440). Work therapy was prescribed by the artist himself as well as by his doctors (letter 489 et al.).

Because of Vincent's exposure to absinthe and turpentine and his subsequent behavior pattern I have suggested[11] that he may have developed a pica (a craving for unnatural articles of food) for terpenes which would explain his reckless use of camphor for insomnia (letter 570), attempt to drink turpentine (letter 590a), and nibbling of mixed pigments (letters T23, T24). This working hypothesis has ramifications for both neurological and gastrointestinal aspects but their discussion will be delayed until chapters 4 and 5.

The Hugo van der Goes connection

The London office of Goupil & Co. was particularly busy with the sale of reproductions when Vincent joined them in 1873. In this manner van Gogh became familiar with at least two works by the Belgian artist Émile Wauters (letter 10). One of these paintings, *The Madness of Hugo van der Goes,* would be mentioned at least twice more in correspondence (letters 514, 556). The references and analogies take on added significance in terms of Vincent's luckless love affairs, his perception of his own illness, and his sojourn in the asylum at St. Rémy.

Hugo van der Goes (*ca.* 1430–1482) was a Flemish painter of considerable merit and success, even in his own lifetime. He was born in Ghent and became a master of portraits. Unfortunately, very few of his pictures have been preserved but a handsome example on wood panel, *Portrait of a Donor with John the Baptist,* is held by the Walters Gallery in Baltimore, and other paintings are on display in Berlin, Vienna, and Florence.

Van der Goes fell in love with the daughter of a wealthy man, who had no regard for artists and denied her permission to marry the painter. She

Figure 3.2. Émile Wauters, *The Madness of Hugo van der Goes*, detail, 1872, Musées Royaux des Beaux-Arts, Brusssels.

retired to a convent in Brussels and Hugo sought solace in the convent of Rouge Cloître, in the Forest of Soigne. The monks afforded him the special privilege of a studio and encouraged him to continue rendering the likenesses of wealthy patrons, who thus helped to support the monastery. The artist was also allowed to take trips, and on the occasion of a return journey from Cologne he was seized with a fit of madness and had to be restrained from killing himself. Supposedly, the monks considered Hugo to be under the influence of an evil spell (à la Saul) from which he might be delivered with a musical rendition (à la David). The story goes that Hugo recovered from that insult but later succumbed to further crises and died in the convent.[12]

It was this musical exorcism that Émile Wauters captured so well in *The Madness of Hugo van der Goes* (Figure 3.2) and for which he received public acclaim at the Brussels Exhibition of 1872. (The picture may have been seen there by Theo van Gogh, in connection with his first professional appointment.) In this life-sized painting two of three monks on the right watch with anxious expectation while a colleague conducts a small group of singers and musicians on the left of the seated Hugo van der Goes. The bearded, central figure with clasped hands, fixed gaze, and monastic robe rivets the viewer. After being displayed in the Paris Salon of 1875, and possibly viewed there by Vincent van Gogh, the painting was purchased by the Belgian government and now resides in the Musées Royaux des Beaux-Arts, Brussels.

Vincent wrote from Arles, on July 29, 1888, "Not only my pictures but I myself have become haggard of late, almost like Hugo van der Goes in the picture by Émile Wauters. Only, having got my whole beard shaved off, I think that I am as much like the very placid priest [monk] in the same picture as like the mad painter so intelligently portrayed" (letter 514). The month before, he had written to Émile Bernard that "after four weeks here [in Arles], ... I had a fit of melancholy like yours" (letter B7) and then in the next letter, "you ought to live like a monk who goes to a brothel once every two weeks – that's what I do myself; it's not very poetic, but I feel it my duty after all to subordinate my life to painting" (letter B8). Van Gogh was seemingly preoccupied with painting in the monastic setting.

In October 1888, Vincent was anticipating the arrival of Gauguin, working hard, and desperately in need of another 50 francs. He painted the following picture in words: "I am not ill, but without the slightest doubt I'd get ill if I did not eat plenty of food and if I did not stop painting for a few days. As a matter of fact, I am again pretty nearly reduced to the madness of Hugo van der Goes in Émile Wauters' picture. And if it were not that I have almost a double nature, that of a monk and that of a painter, as it were, I should have been reduced, and that long ago, completely and utterly, to the aforesaid condition" (letter 556).

Vincent returned again to the van der Goes theme in 1889. While exploring possible alternatives to the Arles hospital, he regretted not being able to find a haven "where they would be willing to take me in for nothing, even supposing I took [upon] myself all the expenses of my painting and left the whole of my work to [them]" (letter 588, April 30, 1889). This was similar to the arrangement of van der Goes with the Convent of Rouge Cloître. And then he did end up in Saint Rémy (which takes its name from the Bishop of Reims who was said to have brought about a miraculous cure there in the year 500) at an asylum that had been the former Priory of Saint Paul de Mausole.[13] With the passage of time van Gogh became irritated by the cloistered features of both the Arles hospital and the St. Rémy asylum, developed "a horror of all religious exaggeration" (letter 605), and eventually felt that his fits had an element of "religious exaltation" about them (letter 607). As far as we know he was never subjected to a musical exorcism!

Gastrointestinal problems

One of the most frequent complaints in Vincent's correspondence concerned his stomach and his digestion. It is worth mentioning that "stomach" in the vernacular means any part of the gastrointestinal system from the stomach itself through the small and large bowels, or influences thereon. The importance of the medical history in approaching the patient with gastrointestinal disease is stressed by Isselbacher and May,[14] who advise that the questioning should include the location of the pain and whether it is circumscribed or diffuse; does it occur before, during, or after eating? (if so – how soon?). The combined occurrence of anorexia and weight loss makes a modern physician suspicious of underlying depression as well as of a malignancy. What is the relation of the "stomach" pain to bowel movements? and so on.[14] Unfortunately, there is no record of these questions having been asked of Vincent.

Accordingly, it is difficult to assess the exact nature, let alone the immediate cause, of the numerous complaints from Vincent about his stomach being "upset" (letter 316), "deteriorated" (letter 448), "terribly weak" (letters 474, 480), "disorder[ed]" (letter B 4), and so forth, together with attempts to "restore the stomach" (letter 321), and follow doctor's orders to "take care of my stomach" (e.g., letters 448, 449). It was such an ongoing problem that Vincent even took it upon himself to record when his stomach and/or digestion were a little better for a change (e.g., letters 458, 520, 530, 569, 607).

References to gastrointestinal complaints

Stomach: 307, 316, 321, 448, 449, 474, 478, 480, 530, 590b, 592, 606, 607, B4, B17, Tralbaut p. 177
Digestion: 215, 442, 450, 458, 469, 492, 520, 569 638, W5
Obstipation: Tralbaut p. 177
Laxatives:
    sour apples, 316
    olives, Johanna Bonger memoir page L
    castor oil, Tralbaut p. 177
Calls for better hygiene: 489, 492, 514, 576

Some of these gastrointestinal complaints were probably secondary to problems with his teeth, throat, availability of nutritious food, alcohol abuse, and so on. Accordingly, there is a fair amount of overlap with some categories which have already been discussed. In particular, it should be noted that regular consumption of absinthe (or any alcoholic beverage for that matter) causes stomach irritation and eventually general upset of the nervous system; this will be discussed in more detail in the next chapter. There are also a number of neurological diseases that can manifest gastrointestinal disorders.

It is worth mentioning that the living conditions in Vincent's domiciles, particularly during the last ten years, could only be described as modest at best. For example, he remarked that there was no lavatory in the Yellow House and that he used one in the hotel next door (letter 480). We can reasonably assume that he often used public facilities. While on his painting trips he undoubtedly relieved himself in the field. I mention this only in the sense that there may have been little opportunity, nor desire for that matter, to observe his urine or feces before or during episodes of illness, and even Vincent may have been reluctant to converse about diarrhea or constipation. There is nothing along these lines to be found in the published letters, which is unfortunate for the retrospective analysis because these aspects can be very instructive.[14] Three references that could be interpreted as a need for laxatives are given in the table. The comment from Tralbaut on castor oil in Antwerp has already been discussed in chapter 2.

Nicotine, caffeine and other chemicals

There are several references to Vincent's proclivity for smoking, he was never without his pipe and even on his death bed Dr. Gachet lit up for him. He acknowledged smoking too much on several occasions. For example, from Antwerp, he admitted to being in terrible shape and noted that, "Now I have made it worse by *smoking a great deal,* which I did the more because then one does not feel an empty stomach so much" (letter 449). I have already quoted from letter 481 in which he referred to drinking and smoking too much at the time he left Paris. In Arles, Vincent referred to heavy smoking, and invoked the excuse of the de Goncourt brothers, "We used to smoke very strong tobacco to stupefy ourselves in the furnace of creation" (letter 507). While in the hospital at Arles, he protested that, "they even forbid me to smoke" (letter 579); it was such a habit that he seemed as much concerned with this restriction as with the local protest (see chapter 2), which had led to his being detained.

When Vincent was prepared to believe that epilepsy was the root cause of his crises he tried very hard to make a case for excusing smoking completely (letter 585). But later when things were going well, he was able to attribute the change to "almost half a year now of absolute *frugality in eating, drinking, smoking*" (letter 595). Apart from the inherent dangers of smoking, we should also note that nicotine lowers the threshold for the expression of symptoms due to other drugs that are taken in concert.[15]

References to drugs and chemicals

Prescribed:
Quinine 206, quinine wine 572
Alum 206
Zinc sulfate (hospital register
   in The Hague)
Potassium iodide 489
Potassium bromide 574

Self-administered:
Camphor 570
Turpentine 590a
Mixed pigments T23, T24
Alcohol (numerous, see
   References to alcohol)

There are several implications for excessive coffee consumption; letter 581, for example, contains Dr. Rey's accusation about Vincent sustaining himself on coffee and alcohol instead of eating properly. Also, van Gogh mentioned to his sister (letter W4) that he had been in the habit of trying to immerse his melancholy in copious amounts of bad coffee, but doubted that it worked. Notwithstanding the well-documented physiological effects of excessive caffeine ingestion, in Vincent's case the problems from this source were probably minor compared with those arising from some other self-administered chemicals (see chapters 4 and 5).

Eye problems

Vincent had two documented episodes. The first occurred in The Hague in early 1883 (as he approached thirty years of age) and may have been referred pain, "These days I am troubled with a bad toothache which sometimes affects *my right eye* and ear; however it may be partly due to nerves" (letter 255). And a short time later he wrote, "If my eyes do not improve, I'll take your advice and bathe them with tea. As it is, they are getting better, so for the present I'll leave them alone" (letter 265). A more serious complaint came in Arles, during the second and third weeks of October, 1888.

It started with letter 552: "I am really so tired that I can see my writing isn't much. ... I am really falling asleep and I can't see any more, *my eyes are so tired.*" It may have been provoked by the prevailing wind: "there is a very violent mistral that raises clouds of dust which whiten the trees on the plain from top to bottom" (letter 553). But the problem persisted: "*My eyes are still tired*" (letter 554) and again, "I do not know what I shall undertake next, for *my eyes are still tired* even yet" (letter 555). And in answer on October 19, Theo commented, "It is very annoying that you have been having *trouble with your eyes.* What can be the cause of that?" (letter T1). It is never resolved for us by the subsequent correspondence.

Many of the letters from Arles during the summer of 1888 mentioned the heat and the intensity of the sun, which "beats down on one's head ... [and] makes one crazy" (letter B15). Vincent's reckless exposure while painting in the fields also received comment (letters 512, B7). Whether heatstroke or some other insult to his system lead to an aberrancy in color vision perception, specifically a xanthopsia, will be addressed in chapter 8.

On January 17, 1889, Vincent wrote, "After my illness [the first major crisis at Christmas 1888] *my eyes have naturally been very sensitive*" (letter 571). This aspect of Vincent's illness and the possible effects on his work remain intriguing. Unfortunately, there are no further clues in subsequent correspondence.

Some symptoms with multiple interpretations

Finally we come to several comments from Vincent which involve serious but rather non-specific symptoms. They are important to any reasonable diagnosis that addresses his total condition, and are listed below without comment for the moment.

References to other symptoms

Malaise, and/or weakness:
148, 207, 234, 238, 302,
304, 305, 308, 309, 440,
442, 449, 452, 462a,
480, 489, 521, 531,
544a, 558, 629, 636,
B15, W11

Fever:
172, 173, 200,
206, 215, 216,
302, 469, 576,
R10

The illness of Theodorus (Theo) van Gogh

In contrast to the wealth of attention that has been paid the artist, comparatively little has been written about the health record of Theo. A unilateral impression can be gleaned from *The Complete Letters* by interpreting Vincent's responses to Theo's comments about his episodes of sickness. So few of Theo's letters have been published that attempts to construct the complete picture are fraught with gaps. The process is made even more frustrating by our limited access to Theo's letters to his wife, and to other members of his family. Furthermore, there is an embargo on access to

Johanna van Gogh-Bonger's diary, which may contain some useful data on Theo's final years. Rewald[16] assembled information on the last six months of Theo's life from letters involving Camille Pissarro and Dr. Paul Gachet. Some previously unpublished letters were also revealed by Hulsker.[17]

It is clear that Theo suffered from at least one debilitating bout of sickness in his youth and experienced severe neurological and psychiatric problems near the end of his life. Theo was certainly grief-stricken by Vincent's suicide, but his own death, by natural causes just 184 days later, was due to an underlying illness which had great similarity to that of Vincent. In fact, Vincent described it as "*our* neurosis ... a fatal inheritance" (letter 481).

Theo was born in Zundert [Holland], May 1, 1857; he was four years and one month junior to Vincent. Little is known about Theo's schoolboy years. At the age fifteen he assumed his first position with the Goupil & Co. Gallery in Brussels, and he worked there for 10.5 months before transfer to The Hague (letter T43). At nineteen, Theo suffered an apparently life-threatening illness which included fevers; he was incapacitated approximately 52 days.[18] A letter within the family fixed the starting date as Tuesday, September 26, 1876, because on that day the young man had to miss work. A short time later his father went to The Hague and attended the patient for a few days; he was relieved by his wife who stayed until October 23, 1876. When sufficiently recovered to make the journey to Etten, Theo joined his family and stayed there until November 16, 1876.

Sister Anna expressed her concern from England and indicated that the crisis itself had lasted about two weeks.[18] The seriousness can be gauged from the father's comment [on October 19], "Thank God who has spared your life." Three months later the parents recalled how Theo had been "dangerously ill and weak." Vincent was out of touch; he made but one reference from London, "I heard from home that you are ill" (letter 75, October 3, 1876).

Theo worked steadily in The Hague until mid-1880 and then was transferred to Paris. He expanded his horizons in that city and was encouraged by Vincent, "one must not hesitate to go to a prostitute occasionally " (letter 173, January 1882). In the same letter Vincent referenced, but did not document, an episode of illness concerning Theo, "I am very sorry to hear that you don't feel well either." Comparative remarks – "[I had] almost three days with fever and nervousness, now and then accompanied by headache and toothache .... sometimes, like now, a heavy depression comes over me, and then it's hell" – provide an inkling of Theo's symptoms. Toward the end of the same year, Vincent made another passing reference, "[I] hope that your headache is not chronic or one which returns again and again. With me it is also more a feeling of uncomfortable fatigue than of excessive pain" (letter 234).

The next available comments on Theo (at age 29) came from his brother-in-law to be, Andries Bonger, to his own parents. Extracts of these letters were reproduced in *The Complete Letters* as item 462a. His friend was obviously concerned and remarked that "Theo is still looking frightfully ill; he literally has no face left at all" (June 23, 1886); and, the next year, "[Theo] stands in great need of being renovated. He continues to look sick and emaciated, and he is feeling weak" (February 18, 1887). A previously unpublished letter[19] from Bonger, on December 31, 1886, mentions, "serious nervous afflictions, so bad that he [Theo] could not move. ... [he is] still stiff, as [one would be] after a fall." Four months later Theo admitted that his own mental illness contributed to current stressful interactions with Vincent: "I was ill, especially mentally."[20]

After Vincent forsook Paris for the South he revealed that his lifestyle during the previous two years had been detrimental to his health and he implied that Theo's was also suboptimal. While Vincent wrote mostly about his own problems there was one letter, in May 1888, which included comments about Theo's: "What you write about your visits to [Dr.] Gruby has distressed me, but all the same I am relieved you went. ... [your] dazedness – the feeling of extreme lassitude ..." [the rest of the letter is about the dubious value of potassium iodide as a tonic and the relative merits of Drs. Gruby and Rivet] (letter 489). In September Vincent wrote, "It isn't cheery news that the pains in your leg have come back (letter 543). In October, Vincent inquired, "How are those sciatica pains, have they stopped?" (letter 550) and again "How are the pains – don't forget to tell me about them" (letter 554). We assume they disappeared but the letters provide no closure.

In the midst of his own concerns at St. Rémy in mid-June 1889, Vincent found time to encourage Theo, "I so often think that after some time your marriage, I hope, will give you back your old vigor, and that a year from now you will be in better health" (letter 595). In July of that year Vincent told his mother, "I am not surprised that he [Theo] is coughing, etc. Sometimes I have wished that they [Johanna and Theo] lived outside Paris" (letter 598). To Johanna he remarked, "As for what you say about Theo's health ... I share your anxiety ... [but] I have realized that his health is like mine, more changeable and uneven than [it is actually] feeble" (letter 599). Johanna was concerned that neither Theo nor herself was in very good health (letter T11). Theo complained that, "the heat has been overwhelming, and I felt so weak that everything made me feel extremely tired" (letter T12, July 16, 1889) and "As for me I look like a corpse but I went to see Rivet who gave me all sorts of drugs which at least do me enough good to put a stop to my cough which was killing me. I think that it is over now" (letter T13, July 29, 1889). And Vincent added, "I'm very glad that Rivet has rid you of your cough, it worried me somewhat too" (letter 602). By mid-August 1889 Theo felt "a lot better than I did some time ago, and my cough is entirely gone now thanks to Rivet's remedies" (letter T14).

During the remaining period that Vincent was at St. Rémy we have little further information on Theo except to note that he appeared to be stressed by the pressures of work, marriage, and fatherhood. Vincent left St. Rémy on May 16, 1890 and he then stayed just three days in Paris with Theo and family. The next month Vincent related to his sister Wil that "Theo's cough was worse than it was when I left him two years ago" (letter W22). (This indicates that Theo's respiratory problem started at least in the beginning of 1888, before Vincent left Paris for Arles.)

In July 1890 Theo took his family to Holland and returned on his own to Paris. Shortly thereafter he received the news that Vincent had shot himself, rushed to his brother's side, and watched him die. There are no reports indicating that Theo was incapacitated, either physically or mentally, at the funeral and during the weeks immediately following Vincent's death. Notwithstanding extreme grief, Theo wrote a beautiful letter[21] about his brother to his mother on August 1. In another lucid and insightful letter to his sister Elisabeth,[22] Theo mentioned his plan to show Vincent's paintings in Paris that October. This last item seems to have escaped Hulsker,[23] who failed to quote the letter in its entirety as claimed. Theo returned to Holland on August 3. His whole family was back in Paris again by mid-August.[16]

With the assistance of Émile Bernard, Theo was able to arrange a display of Vincent's paintings in the rooms of their new apartment. Many friends, including the critic Albert Aurier, were excited by the exhibition. According to Bernard[24] Theo was noticeably agitated shortly thereafter and "lost his reason and collapsed; he was paralyzed." The coughing spells returned and were successfully treated with an unspecified medication from a Dr. Van der Maaten.[25] On September 12, 1890, Theo wrote Dr. Gachet, "I still do not feel good about my health; my head is swimming and writing makes me a bit dizzy. My nerves still have the upper hand."[16,26] To his sister Wil[25] Theo described how, "the drops [of medicine] of that Dr. [Van der Maaten] ... helped to stun me during the night and prevented me from coughing, but they gave me hallucinations and nightmares ... I was literally crazy." He gave up on the cough medicine.

On October 10, 1890, Andries Bonger wrote to Dr. Gachet: "Since yesterday my brother-in-law van Gogh has been in such a state of overexcitement that we are seriously worried. If it were possible we would be extremely grateful if you would go to see him tomorrow, making believe that you are merely dropping in. Everything irritates him and makes him lose control. The excitability has been caused by a dispute with his employers, as a result of which he wishes to set up his own business without delay. The memory of his brother haunts him to such a degree that he is angry at all those who do not go along with his ideas. My sister is at the end of her strength and does not know what to do. I hope that it will be possible for you to come; if not please have the kindness to write me a note of advice."[16,27] Before Gachet could respond, Theo was hospitalized [on October 12, at Maison Dubois] and two days later he was transferred to the private clinic of Dr. Blanche in a suburb of Paris.[28] The attending physician was Dr. Antoine Blanche, son of the clinic's founder, and it is worth noting en passant that a grandson had purchased an important Degas painting from Theo van Gogh in 1888.[*]

On October 18, 1890, Camille Pissarro wrote his son, "It appears that Theo van Gogh was ill before his madness; he had uremia. For a week he was unable to urinate; added to that were the worries, the sadness, and a violent argument with his employers concerning a Decamps painting. As a result of all this, in a moment of exasperation he thanked the Boussods and suddenly went mad. He wanted to rent [the Café] Le Tambourin in order to form an association of artists. Finally he became violent. He who had so loved his wife and child wanted to kill them. In brief they had to put him into Dr. Blanche's sanatorium."[16,30] In the meantime, a distant victim was Paul Gauguin, who had received a cable in Brittany – "Departure for tropics assured, money follows, Theo, Director"[31] – only to have his hopes dashed when he learnt about the unsound mind of Theo.

Mr. Tersteeg, who had arrived from Holland to discuss Theo's position with the Boussod & Valadon management, and Dr. Gachet observed Theo in the Blanche sanatorium but were unable to converse with him. According to a letter written by Andries Bonger, Dr. Rivet suggested that "his [Theo's] case [was] far worse than Vincent's, and that there [was] not a spark of hope."[32] On November 14, 1890, Camille Pissarro informed his son[16,33] that Andries Bonger now considered Theo to be calmer and that it would be possible to take him to Holland.

Dr. Frederik van Eeden[**] of Holland visited Theo in Paris and found him

---

[*]   According to ledger sheets reproduced by John Rewald,[1] *Foyer de la danse*, was purchased by Jacques-Émile Blanche on November 29, 1888, for 8,000 fr. This represented a profit of 2,780 fr. for Boussod & Valadon. On October 12, 1889, Theo van Gogh sold a Monet painting for 10,350 francs, a record at the time.[29] Theo had more than a modicum of ability in selling art.

[**]   Van Eeden (1860–1932), a poet, novelist, and playwright as well as a psychiatrist, was mentioned by Vincent in connection with his painting *La Berceuse*, "our lullaby or the woman rocking the cradle." A similarity in style with the doctor's prose was noted (letter 571a). The doctor was a founder (1885), and editorial board member, of *De Nieuwe Gids* (The New Guide).

"much weakened, both mentally and physically."[34] On November 17, Theo came in a straight-jacket to Willem Arntsz Stichting, the mental institution in Den Dolder, just outside Utrecht. Based upon the legally prescribed declaration [which mentions inheritance, chronic illness, and excessive work and sorrow, as causes of the psychotic illness], the judge of the second district of Amsterdam had decided to allow Theo's admission into the mental home. From the first day of internment Theo had "no notion at all of place or time – he mutters incoherent words" [from the notes of Dr. Van der Eeden]. He died in Willem Arntsz Stichting on January 25, 1891. The final diagnosis according to local records[34] was dementia paralytica (parenchymatous neurosyphilis).

The illness of Willemina [Wil] van Gogh

Vincent wrote at least 23 letters to his youngest sister, Wil, but none of her letters to him are available. Documentation of her health is scant although Vincent gives us some indication that Wil had bouts of nervous troubles. While writing from Arles in mid-1888, he seemed to include her in the family propensity for a *nervous temperament* and he suggested that "The remedy for the immersion [in melancholy] which you mention is not, as far as I know, to be found growing among the herbs with healing powers. ... I am in the habit of taking large quantities of bad coffee in such cases ... because my strong imaginative powers enable me to have a devout faith ... in the exhilarating influence of said fluid" (letter W4). Later in the same letter he remarked, "I am extremely glad that you [Wil, age 26] have regained your health" although the nature of her illness is never defined. About a month later he asked, "How is your health? Good I hope. Above all you should try to get into the country as much as possible. Here I am afflicted now and then with an *inability to eat,* something *of the sort you suffered from at the time*" (letter W5).

On May 2, 1889, Vincent was concerned about the impending costs at St. Rémy and worried about future housekeeping costs that Theo would be faced with in Paris. He went on to say, "in the future there might possibly be our sister to be provided for" (letter 589). We do not know the full circumstances. Perhaps the brothers were simply concerned that Wil had not found a husband; Vincent had earlier suggested that Theo arrange a meeting between Wil and their Belgian friend Eugène Boch (letter 532). We do get the impression that Vincent had a protective concern for his youngest sister, the only one of his sisters with whom he had any affinity after childhood.

In January 1890, from St. Rémy, Vincent wrote to his sister and asked if she had suffered from influenza. He said that he had observed such a patient in the asylum who had some "rather disquieting nervous complication with distressing climacteric [the syndrome associated with menopause]" (letter W19). He hoped she would write soon and that she had recovered from her illness. Once again the nature of this malady is not resolved in the letters.

Wil lived with her mother in Leyden for many years. According to Hulsker[17] she showed signs of "insanity" at the age of 41 and was committed to an asylum in The Hague. At the end of 1902 she was accommodated at Veldwijk, a mental home in Ermelo, run by the church. She died there in 1941 having spent the latter half of her 79 years in a mental asylum.

Summary

Vincent van Gogh's ailment was characterized by episodes of acute mental derangement and disability which were separated by intervals of lucidity and creativity. The artist himself as well as attending physicians, relatives, and friends were all surprised and encouraged by the rapidity of the recoveries after each crisis. His serious illness developed late in the third and played into the fourth decade. There were frequent gastrointestinal complaints and at least one bout of constipation that required medical intervention. Hallucinations, both auditory and visual, and nightmares were associated with the crises. The periods of incapacitating depression and physical discomfort were severe and grave enough to provoke self-mutilation and eventual suicide.

There was a family history of mental illness which affected Vincent, his brother Theo, and sister Wil, i.e. three of the six siblings. The illnesses of Uncles Cent and Hein may also bear some relationship with those of the younger family members. The hereditary aspect is important to the understanding of the artist's case.

Vincent's symptoms were palliated during institutionalization due to a better diet, abstinence from alcohol, and administration of bromide therapy. In spite of the severity of crises he did not experience any permanent, functional disability which affected his painting or his writing. There were many exacerbation factors. These included malnutrition and fasting, environmental exposure, and excessive ingestion of alcohol. Alcohol problems were compounded by a proclivity to drink absinthe, a liqueur containing a toxic compound, thujone, and several related terpenes. Absinthe may have exacerbated van Gogh's underlying illness and also induced a pica for terpenes, the documented examples being camphor and pinene (in turpentine).

References and notes

1. du Quesne-van Gogh E. 1913. *Personal recollections of Vincent van Gogh.* (translated by KS Dreier) Boston & New York: Houghton Mifflin Company.
2. van Gogh-Bonger J. 1978. Memoir of Vincent van Gogh. vol I, XV–LIII, in: *The Complete Letters of Vincent van Gogh.* 2nd ed. Boston: New York Graphic Society.
3. Holmes KK. 1987. Gonococcal infections. vol 1, chapter 104, in: *Harrison's Principles of Internal Medicine.* 11th ed. E Braunwald et al. (eds.). New York: McGraw Hill Book Co.
4. Devine CJ Jr. 1986. Surgery of the urethra. vol 3, p 2865, in: *Campbell's Urology.* 5th ed. PC Walsh et al. (eds.). Philadelphia: W.B. Saunders Co.
5. Holmes KK & Lukehart SA. 1987. Syphilis. vol 1, chapter 122, in: *Harrison's Principles of Internal Medicine.* 11th ed. E Braunwald et al. (eds.). New York: McGraw Hill Book Co.
6. Tralbaut ME. 1981. *Vincent van Gogh.* pp 177–8. New York: The Alpine Fine Arts Collection Ltd.
7. van der Wolk J. 1987. *The Seven Sketchbooks of Vincent van Gogh: a Facsimile Edition.* pp 104–5. (translation by C Swan) New York: Harry Abrams Inc.
8. Stein SA. 1986. *Van Gogh: a Retrospective.* pp 132–134. New York: Park Lane.
9. Tralbaut ME. 1981. *Vincent van Gogh.* pp 269–270. New York: Alpine Fine Arts.
10. Arnold WN. 1989. Absinthe. *Scientific American* **260**: 112–117.
11. Arnold WN. 1988. Vincent van Gogh and the thujone connection. *Journal of the American Medical Association* **260**: 3042–3044.
12. Anonymous. 1881. *The Magazine of Art* (London, Paris, New York) **4**: 168–169.
13. Pickvance R. 1986. *Van Gogh in Saint-Rémy and Auvers.* New York: Harry N. Abrams Inc.
14. Isselbacher KJ & May RJ. 1987. Approach to the patient with gastrointestinal disease. vol 2, chapter 232, in: *Harrison's Principles of Internal Medicine.* 11th ed. E Braunwald et al. (eds.). New York: McGraw Hill Book Co.
15. Sollmann T. 1948. *A Manual of Pharmacology and its Applications to Therapeutics and Toxicology.* 7th ed. Philadelphia: WB Saunders Co.
16. Rewald J. 1986. Theo van Gogh as an art dealer. pp 7–115, in: *Studies in Postimpressionism.* I Gordon & F Weitzenhoffer (eds.). New York: Harry Abrams Inc.
17. Hulsker J. 1990. *Vincent and Theo van Gogh: A Dual Biography.* Ann Arbor: Fuller Publications.
18. Hulsker J. *op.cit.* p 34.
19. Hulsker J. *op. cit.* p 244.
20. Hulsker J. *op. cit.* p 246.
21. Hulsker J. *op.cit.* p 449.
22. du Quesne-van Gogh E. 1913. *Personal Recollections of Vincent van Gogh.* pp 52–53. (translated by KS Dreier) Boston & New York: Houghton Mifflin Company.
23. Hulsker J. *op. cit.* p 450.
24. Bernard É. 1911. *Lettres de Vincent van Gogh à Émile Bernard.* Paris: Ambroise Vollard.
25. Hulsker J. *op. cit.* p 452.
26. Gachet P[L]. 1957. *Lettres Impressionnistes au Dr. Gachet et à Murer.* p 153. Paris: Éditions des Musées Nationaux.
27. Gachet P[L]. 1956. *Deux Amis des Impressionnistes: Le Docteur Gachet et Murer.* pp 125–6. Paris: Éditions des Musées Nationaux.

28. Gachet P[L]. 1957. Les Médecins de Théodore et de Vincent van Gogh. *Æsculape* **40**: 2–37.
29. Moffet CS & Wood JN. 1978. *Monet's Years at Giverny: Beyond Impressionism.* p 155. New York: The Metropolitan Museum of Art.
30. Letter from Camille to Lucien Pissarro, October 18, 1890, *Lettres.* pp 188–9. This letter appears only in the French edition according to Rewald (ref. 16).
31. Alexandre A. 1930. *Paul Gauguin: Sa Vie et le Sens de son Oeuvre.* p 108. Paris, as quoted by Rewald (ref. 16).
32. Hulsker J. *op.cit.* p 454.
33. Letter from Camille to Lucien Pissarro, November 1890, *Lettres* (French edition), p 189, as quoted by Rewald (ref. 16).
34. This item and subsequent details of Theo van Gogh's illness after Paris were supplied by Dr. A. Pietersma, Archief Dienst Gemeente, Utrecht.

Chapter 4
Absinthe

Figure 4.1. Henri de Toulouse-Lautrec: *Portrait of Vincent van Gogh*, 1887, Paris, pastel on paper, 54 × 47 cm (21.3 × 18.5″),
Vincent van Gogh Stiftung/Van Gogh Museum, Amsterdam

> *If there were any difference in the composition*
> *[of paints] that's what would interest me more.*
> *And in paint there is adulteration just as there*
> *is in wines, and how can you judge accurately*
> *when you, like me, do not know chemistry?*
> Vincent to Theo, letter 642, from
> Auvers-sur-Oise, June 17, 1890.

Absinthe was in vogue during the nineteenth century and the beginning of ours. It was a clear green from the bottle and then, upon gentle dilution with water, displayed a beautiful pale-yellow opalescence. Artists painted and poets personified; men and women embraced the ritual of presentation as well as the appearance, taste, and excitement of this alcoholic drink. Some of the most creative people of the nineteenth century were included. Were they exposed to new ideas, exotic experiences, unique feelings, (a different view of a "Night Café"), perceptions that were otherwise unrealizable? Unfortunately, too much absinthe wracked your brains, exacted a price, and shortened your life.

The aesthetics of absinthe drinking contributed to its popularity. Nevertheless, one looks to the mood-altering chemicals that were unique to this liquor in order to rationalize the volumes consumed in some quarters, and the fifteen-fold per capita increase in France from 1875 to 1913, when the annual consumption attained a massive 368,400 hectolitres of absinthe (about 9.7 million U.S. gallons). As one of my practical colleagues has remarked, "whenever you have a whole nation interested in imbibing a particular beverage, there must be more to it than poetry and attractive colors." In the department of Bouches-du-Rhône, which includes van Gogh's southern venues of Arles and St. Rémy, the annual consumption was an impressive 2.45 litres per head, which was more than 4 times the national average.[1]

In the latter half of nineteenth century France, *l'heure verte* (the green [cocktail] hour) was an established daily event, and some Parisian clubs were dedicated to the drink. The "Old Absinthe House" in New Orleans, Louisiana, dates from the same era. As late as 1914, Pablo Picasso constructed a set of six abstract sculptures, in enamelled bronze, depicting absinthe glasses and incorporating authentic silver absinthe spoons. The artist's work, apparently in response to contemporaneous legislative attempts to ban absinthe in France, has been interpreted[2] as Picasso's endorsement of freedom of choice in matters alcoholic. Between 1905 and 1913, Belgium, Switzerland, the United States, and Italy all banned absinthe, and by 1915 the French government made absinthe less available, but it was not effectively banned there until somewhat later.

In the last chapter I referred to Vincent's taste for absinthe and quoted Signac, "[Vincent] would take his seat on the terrace of a café. And the *absinthes and brandies* would follow each other in *quick succession*" (letter item A16). One commentator was silly enough to challenge Signac's memory, and wondered if Vincent ever touched absinthe. The drink was so popular in Paris and Arles that it seems to me more reasonable to put it the other way round; why would Vincent not be drinking absinthe when everybody else was! On the other hand, we all wonder just how much absinthe Vincent consumed, and what was his individual reaction to it.

Absinthe fits

As early as 1859, Auguste Motet wrote a medical dissertation[3] on precocious delirium after absinthe abuse, and on epileptic attacks accompanying absinthism.* A formal recognition of the toxicity of absinthe, something that was long suspected by common observation, was thus scientifically stated, but by the nature of this document it reached a limited audience. Probably the earliest journal report, 1864, was by Louis Marcé[4] in the form of a short note (introduced by no less than Claude Bernard) on the "double action" (something in addition to alcohol) of absinthe in causing convulsions, involuntary evacuations, and other alarming symptoms in experimental dogs and rabbits. Dr. Marcé did his research at the Hôpital Bicêtre in Paris. The investigations were continued there, and later at Saint Anne Asylum, by his student Valentin Magnan. Both institutions were renowned for the treatment of the insane.

The results of Magnan's investigations on animals, and his observations on absinthe addicts, were the most influential of the medical data that finally lead to interdiction by the French and other governments. The special action of absinthe, as compared with other drinks such as brandy, whisky, gin, and rum (which are also high in alcohol) was soon realized as due to the essential oils that were integral to absinthe.[5,6] Thus, according to Magnan, "Absinthe ... can rapidly give rise to hallucinations and delirium before the alcohol contained in the liqueur has had time to produce [even a] trembling in man ... [whereas usually] with acute alcoholism the trembling and intellectual disturbances progress simultaneously, or present only

---

* Absinthism is defined in *Dictionnaire de Médecine*, Littré & Robin, 1865, as a "variety of alcoholism, acute or chronic, caused by abuse of absinthe, and leading (more frequently than alcoholism) to mania and softening of the brain. ... [This is] due to the poisonous action of the essences which enter into the composition of this dangerous liqueur."

slight differences in their successive evolution."[6] Absinthism was distinguished by sudden delirium, epileptic attacks, vertigo, hallucinatory delirium more active and more impulsive than with alcohol alone, and sometimes unconsciousness; the patient sometimes exhibited a thickness and embarrassment of speech, and inequality of the pupils.[7]

A typical case of absinthism

Dr. Amory, an American student of Magnan reported on human and animal experimentation in 1868, in the forerunner[8] of *The New England Journal of Medicine*. The following is a précis of one of their cases.

A grocer in Paris was unfortunate in business and turned to the retail sale of wines, but soon assumed the habit of taking a glass of cognac with his customers. He did this for two years and developed enfeebled digestion and eventual dyspepsia, with great pain in his stomach after eating. He began to drink absinthe and found, for a short time at least, an agreeable stimulus to his digestive system, but the mornings after were often attended with vomiting and great trembling. His sleep became disturbed by dreams and restlessness. A short time after starting the absinthe habit he took his customary five or six glasses of the liqueur during the day and then attended a funeral. During that ceremony he was seized suddenly with a fit, lost consciousness, fell down, bit his tongue, grimaced, agitated his extremities, and started foaming at the mouth.

He recovered but did not give up his habit, even after several experiences of vertigo and other disturbing symptoms. After another attack he was admitted to the Hôpital Bicêtre and spent an uneasy night in a straight jacket. His expression was leaden, the face covered with sweat, tongue torn at the borders, belly hard with constipation, urine reddish and laden with albumin. The patient exhibited a great deal of trembling, hesitated in his speech pattern, allowed his eyes to wander, was incoherent, had no idea where he was, then imagined he saw his wife and children and even warned them of an impending danger, turned to see rats and spiders at the foot of his bed, observed imaginary flames, expressed fright, and passed out. Laudanum [a solution of opium (morphine) in alcohol] was prescribed.

The hallucinations continued and the patient complained of a pain in each side of his chest, and of abnormal frequency of respiration. After 21 days and an amelioration of all symptoms he was dismissed. Almost seven months later the patient was brought back to the hospital and admitted to still drinking absinthe but having cut back on brandy. In the interim he had been intoxicated several times, wherein he was violent towards his own family. He had even more hallucinations than before, experienced pain in the abdomen in the region of the stomach, but had no constipation this time. Two days later he developed pulmonary trouble with abnormal sounds [râles] accompanying the respiration. By the sixth day he was no longer hallucinating or having nightmares, and the secretion of protein in the urine had subsided. After another fortnight the patient was dismissed for convalescence but was never seen or heard from again.

The sudden attacks, including distortion of features, convulsions, tongue biting, foaming at the mouth, and complete loss of consciousness were described as resembling epilepsy, but there was no evidence of hereditary epileptiform convulsions and no accident in which the cranium had been traumatized. The patient was free of epilepsy during both hospitalizations. Also, there were no attacks before he started drinking absinthe at the age of 32, nor during the previous two years devoted to brandy. On the basis of many such cases Magnan implicated the plant oils in absinthe and tested the essences of wormwood and anise (major components). In experiments with dogs he found that anise was well tolerated, and there was not the slightest symptom of a convulsion. Wormwood oil was more difficult to administer because it caused vomiting and diarrhea but when encapsulated in gelatine remarkable symptoms were provoked.

"In a weak dose, essence of absinthe induces a more or less marked muscular tremor, little abrupt jerking shocks ... in the muscles of the [dog's] neck ... [and the clonic] contractions reach in succession the muscles of the shoulders and the back and then provoke stronger abrupt [tonic] contractions."[6] Magnan emphasized that absinthe provoked spasmodic contractions in the forequarters before producing generalized convulsions, whereas alcohol produced a paralysis of the hind quarters and then progressed to the fore part of the body of the experimental animal. In a larger dose, "epileptic attacks come on: the animal loses consciousness, falls, and stiffens in tonic convulsions which form the first stage of the fit. ... In the interval between two epileptic attacks, and sometimes before the convulsive symptoms, or even without convulsions, the animal is seized with an attack of delirium."[6] The hallucinations in dogs were revealed by their posturing towards a blank wall as if confronting a foe: "[The dog] barks furiously; advances and retires as [if] before an enemy; with an open mouth, he throws his head suddenly forwards, and immediately shuts his jaws and shakes them from side to side as if wishing to tear his prey to pieces. This attack of delirium may occur several times; then the effects pass off, and the animal becomes quite calm."[6] The investigations were extended to the cat, rabbit, and guinea pig. The latter became the experimental animal of choice because it did not vomit, and the effects of absinthe were more easily produced.[8]

The scientific warning from these studies reached the popular press but was matched by strong denials. Government taxes as well as industrial profits were sizeable so that it was not surprising to see opposition to attempts at restricting the sale of absinthe. Consumer reaction fin de siècle varied from mild restraint to complete disdain. Men and women from all walks of life, more or less caught up in a brisk industrial revolution, strove to convince themselves that the risks were small and at least commensurate with the pleasure.

The question of dosage is important. The liqueur contained about 3% essential oils;[1] Marcé used 2–3 grams of absinthe essence per dog;[4] Amory[8] (in Magnan's laboratory) used 4 grams per dog for the more pronounced effects; this represented 67–133 ml of absinthe per animal. Amory[8] mentions an experiment with a guinea pig in which convulsions were induced with 3 ml of neat absinthe. Obviously absinthe contained toxic substances, but needless to say a normal drink of absinthe did not evoke a seizure in a healthy man. Those who supported the drink tried to ignore the demonstrable dangers by invoking an ambivalence about the dose. One of the most amusing of these was "L'Absinthe et le Cobaye" (Absinthe and the Guinea Pig), a tongue-in-cheek poem written in 1901 by Raoul Ponchon,[9] bon vivant and commentator on all elements of Parisian life. The preamble, first, second, fourth and sixth stanzas give the sense; my translation makes no attempt at rhyme.

Absinthe and the guinea pig

> Monsieur Bordas, deputy chief of the municipal laboratory injected 10 c.c. of absinthe into a guinea pig in order to demonstrate the mortal toxicity of this beverage.

Ten cubic centimetres! What a binge!
Why not thirty all at once?
Poor guinea pig! For him the end
is to serve the experiments
of these men of science,
along with his brother the rabbit.

But, good expert whom I respect,
you know all too well
that I inject myself with relatively less.
If I did it your way it would be
as if I should take upon my rascally self
a drink of one litre at one sitting ... Merci!

Besides, what does the thing prove!
That the little guinea pig in question
is not bigger than the force
peculiar to the absinthe?
That it should turn out to be infinitely toxic
to him is not the point for me.

In the same way, my throat
jumps when I try to swallow a sword
(as I have seen them do in Barnum's Circus).
Indeed I do not know for which creature
the sword is ordinary fodder.
How will you? ... to each his own.

More poets and writers

The first of many great alcoholic American writers was Edgar Allan Poe and apparently "absinthe explains Poe's agitation, hallucinations, and deliriums better than opium."[10] For example, his fits occurred in the middle of his drinking and also when he was acutely intoxicated. This suggests the delirium induced by absinthe rather than delirium tremens, which does not occur while the person is drinking, but after the drinking stops, i.e. a withdrawal phenomenon. Poe was a freshman at the University of Virginia for eight months and was remembered by another student as "always seizing the tempting glass, generally unmixed with sugar or water,"[10] which sounds to me like neat absinthe. In later years, Poe's repeated binges on brandy and absinthe were well documented. In his final year (1849) Poe complained of headaches, became depressed and feverish, went missing for five days, showed up in Baltimore, and died in a hospital four days later. He was delirious till the end and talked to imaginary objects, he had lived only 40 years.

Other poets attested to the charms of absinthe and spent a lot of time drinking it. Alfred de Musset reputedly shortened his life with absinthe. Charles Baudelaire, a close friend of Édouard Manet, included absinthe among his vices but got an exaggerated rap for the oft quoted, "be drunk, always" because he actually added some lines later, "with wine, with poetry, or with virtue, as you please," a wholesome choice by any standard! Paul Verlaine could be rampant, coarse, and sensual all in the same verse and was in the habit of awaiting inspiration over a glass of absinthe. Arthur Rimbaud abused it in his youth; his poetic career was finished at twenty and he was in his cups most of that brief time; he died at age 37.

The Englishman Ernest Dowson was a drinker and a writer of prose poems, in that order. He was reasonably accomplished for his short span of 33 years but is best known for his double pun[11] upon the aphrodisiacal rumor, "I understand that *absinthe* makes the *tart* grow fonder."* Dowson popped back and forth between London and Paris and sometimes had difficulty remembering the trips. On the other side of the channel, Alfred Jarry insisted that rational intelligence was quite inferior to hallucinations, and employed absinthe to induce them. His enigmatic, surrealist plays influenced the poet Guillaume Apollinaire, friend of Picasso and Gertrude Stein, and all three enjoyed their absinthes in the bistros of the rue de Seine.

Artists

Les Déclassés (The social rejects), an 1881 painting by Jean-Francois Raffaëlli was a hit in the sixth exhibition of the Impressionists. It depicts two mellow fellows with their opalescent glasses and now goes under the title *The Absinthe Drinkers*. The critic Albert Wolff[12] opined that Raffaëlli, "shows the modest people of Paris as they are – more often than not stupefied by life's hardships." An earlier and more challenging image was Édouard Manet's *The Absinthe Drinker* for which the model was a derelict, sketched outside the Louvre. The unusual choice of subject made it a landmark by today's standards, but the painting was rejected by the 1859 Salon. Equally well known, and offering even a stronger social comment, is Edgar Degas' *L'Absinthe*, 1876, which depicts an intoxicated woman and her withdrawn companion in the café. The graphic work of Honoré Daumier included two lithographs from 1863 with the sarcastic titles, *Absinthe ... the first glass ... the sixth glass,* showing the progression within two drinkers, and *Beer – never ... it takes absinthe to revive a man* as the caption to an inebriated and seedy fellow stating his preference to a more robust companion.

---

*   There is no evidence for absinthe being an aphrodisiac. In common with other alcoholic beverages, it increases the sexual ambition but lowers the performance.

Henri de Toulouse-Lautrec's health was affected by absinthe and he was temporarily institutionalized after a bad bout in March 1899. Lautrec has the dubious distinction of being credited with introducing Vincent to absinthe in Paris (1886–88). Between 1885 and 1892 the consumption of all forms of alcohol in Paris increased but 5%; however, absinthe consumption increased 125%, a measure of its increasing popularity.[13] Lautrec featured absinthe several times in his paintings, including *Absinthe Drinker* of 1888 and *Monsieur Boileau at the Café* from 1893. The latter is worth noting for the full glass of *green* liquid and probably depicts Lautrec's own invention "an earthquake" of absinthe diluted with cognac (rather than water). He also made a pastel in 1887 of Vincent van Gogh sitting before a glass of absinthe.[*] The next year van Gogh himself did a still-life of an absinthe glass plus a decanter of water. Vincent's *The Night Café* includes absinthe drinkers as does Paul Gauguin's rendition of the same room; both done in 1888, in Arles, a veritable center for absinthe drinking.

Harper's Weekly in the U.S. published an engraving of imbibers "who sit at the little marble tables, drink absinthe, and are invariably [militarily] decorated" (April, 1889). The year that Vincent died, the Norwegian expressionist Edvard Munch drew *The Absinthe Drinkers* with pastel and crayon. Pablo Picasso created at least six paintings with absinthe motifs; they range from *Woman drinking Absinthe*, 1901, to the cubistic *Bottle of Pernod and Glass*, 1912. A fine collection of absinthe art, including several of the examples mentioned above, has been assembled by Conrad.[16]

Public health and legislation

After the basic investigations of Magnan and others had gained acceptance, public health workers in France began to collect statistics on absinthe

---

[*] In an earlier publication I used the expression "partaking of a glass." Hulsker[14] objected to "partaking" and insisted on the static message that Vincent only sits before the glass. That Toulouse-Lautrec chose to depict Vincent with a glass of absinthe suggests to me, and to everybody else whom I have asked, that it was a common enough circumstance, and that Vincent drank absinthe. We feel that van Gogh was not in the habit of simply decorating his table "with a glass of absinthe in front of him" as Hulsker implied. That commentator maintains the isolated position that there is no evidence that van Gogh was fond of absinthe, and he also denies all the statements and anecdotes about his drinking problem. Alas, Hulsker defeats his own hypothesis in several places, not least of which when he suggests that the lack of recall of the ear-cutting episode was "because drinking had caused him [Vincent] to black out."[15] Please see chapter 3 for a long list of letter references relating to alcohol abuse.

consumption by département (legislative region). Positive correlations were noted between per capita absinthe consumption and the incidences of neurological disorders, rejections of army conscripts due to psychoses, and stillbirths.[1,7] Some attempts were made to blame heinous crimes on absinthe intoxication. Criminals occasionally gained a measure of sympathy by claiming an absinthe addiction as their court defense. One celebrated case in Switzerland was supposedly the immediate reason for legislative discussion, which resulted in a sweeping referendum in support of absinthe prohibition in July 1908, and a final enactment in October 1910.

Foremost among French officials to champion interdiction on absinthe was Henri Schmidt, an apothecary and member of the Chambre des Députés, who issued a final report[1] in 1915. He analyzed a total of 9,944 patients (71% male) institutionalized during 1907, whose mental derangement was attributable to alcohol. A clear preference in type of alcoholic beverage was obtained from 75% of the men and 60% of the women, some of the results are summarized in the table. Further evidence of the significant contribution of absinthe came from a comparison of the relative values of the so-called "coefficient d'aliénation mentale" (coefficient of mental derangement), Table 2. This value is the ratio of incidence to amount of (absolute) alcohol consumed, for each type of drink.[1] Schmidt's closing line was firm, "The interdiction of absinthe is an enactment of national protection."

Table 2. Preference among alcoholic lunatics (after Schmidt[1])

Beverage	Males (%)	Females (%)	Normalized Coefficient (of mental derangement)
Beer	0.3	0.03	1
Wine	8.1	10.7	4
Cider	1.8	1.7	32
Spirits	38.6	34.5	77
Liqueurs	3.2	5.6	143
Apéritifs	2.2	0.7	170
Absinthe	20.5	6.2	246

Prohibition of absinthe in France was legislated March 1915, more than seven months after the commencement of World War I. Cartoons compared the dangers of absinthe stills with German cannons. One commentator suggested that French national leaders finally realized that in order to save their country everything that made for inefficiency, including absinthe, must be swept away.

The English medical journal, *The Lancet*, from the 1870's through the early 1900's published several short articles on absinthe and absinthe abuse, mostly about "our neighbours the French,"[17] the volumes of alcohol consumed,[18,19,20] and "the absinthe evil."[21] They were also impressed by the number of liquor stores in France; for example in 1874 there were about 383,000 of which 10% were in Paris[18] (the number of liquor retailers matched the bakers and butchers combined),[19] and the number increased about 1% each year for the next 12 years.* This righteous English position was slightly shaken by Brasher,[13] who remarked in 1930 on the prevalence of English cocktails spiked with absinthe essence. In several London clubs of that time it was fashionable to order mixed drinks with a "spot" or a "kick" of absinthe. The sale of absinthe was still unrestricted in England and for the period 1921–29, over 14,000 gallons were imported from Holland, Spain, and France. Somebody was still making it!

One the earliest public health reports in the United States, dated 1868, claimed that the French army had been drinking absinthe of the cheapest quality and that the officers were attempting to restrict it.[23] It is worth considering how unsuccessful they may have been, and whether the habit contributed to France's poor showing in the Franco-Prussian War (foolishly declared by Napoleon III in 1870) and their defeat in less than two months. A longer article on absinthe was offered in 1906 by Walker[24] who similarly indicated that the problem was primarily a French one, but also suggested a growing concern in the U.S. She reminded readers of the differences between absinthe and whisky, described the pathology of absinthism, and summarized, "Absintheurs are restless at night. They suffer from nightmare, nausea, lack of appetite, vomiting, mental dullness and sometimes delirium or mania [depending on the dose]. ... Instead of the simple muscular tremor of delirium tremens, as seen in the [common] alcoholic, the epileptic fit is seen in the absinthe drinker. The fit recurs from time to time." The following year the U.S. Department of Agriculture was asked to measure the extent of cultivation of the requisite herbs and the volume of absinthe manufacture in the United States. The figures were deemed significant and the product dangerous; five years later absinthe was banned in the U.S. The cultivation of wormwood as a cash crop (previously significant in France, Switzerland, Germany, Denmark, Roumania and the United States) was severely curtailed about this time.

---

*   Delahaye[22] essentially agrees with these figures. She says that the 22% increase between 1880 and 1899 was in part due to a relaxed law of 1880 on the licensing of retail shops. This act precipitated 10,000 new shops in six months. In 1909 there was a shop for every 30 adult males in France. Paris had 30,000 compared with 5,860 for London and 5,740 for Chicago.

The herbs of absinthe

The plant products in absinthe varied among manufacturers because of individual recipes, relative costs of the herbs, geographic and yearly variations in crop availability and size, and regional taste. The only universal components of absinthe were alcohol and wormwood essence. In Table 3, I have chosen to list as primary the herbs that went into *Absinthe Suisse de Pontalier,* [25] one of the finer types of absinthe, and one of the simplest and most consistent preparations.[*] The secondary herbs showed the most variation, ranging from zero in some varieties to concentrations which sometimes matched the primary constituents.

Table 3. Plants used in the preparation of absinthe

Plant (parts used)	Alternate names (language)	Linnean genus & species [family]
Primary:		
wormwood (dried leaves & floral heads)	absinthe grande (Fr.), ascenzo (It.) Wermut (Ger.), ajenjo (Sp.)	*Artemisia absinthium* [Compositae]
Roman wormwood (dried leaves & floral heads)	absinthe petite (Fr.) Roemischer Wermut (Ger.)	*Artemisia pontica* [Compositae]
anise (fruit)	anis vert (Fr.)	*Pimpinella anisum* [Umbelliferae]
fennel (fruit)	fenouil de Florence (Fr.)	*Foeniculum vulgare* [Umbelliferae]
hyssop (dried tips & flowers)	hysope (Fr.)	*Hyssopus officinalis* [Labiatae]
lemon balm (dried leaves)	mélisse citronée (Fr.)	*Melissa officinalis* [Labiatae]

---

[*]   The fabrication of so-called "Swiss" absinthe in France was centered at Pontalier, Montpellier, and Lyon, each with regional recipes. Pontalier is a relatively small town in the Jura Mountains, near the border with Switzerland. In its heyday, circa 1905, Pontalier had less than 9,000 people, but 25 absinthe manufacturers with 151 large stills, and a production of over 70,000 hectolitres per year. The largest house was Pernod Fils with more than half of the town's total production.[22]

114

Secondary:		
Chinese star anise (fruit)	badiane (Fr.)	*Illicium verum* [Magnoliaceae]
coriander (fruit)	coriandre (Fr.)	*Coriandrum sativum* [Umbelliferae]
angelica (seeds)	angélique (Fr.)	*Angelica archangelica* [Umbelliferae]
dittany of Crete (dried leaves)	dictame (Fr.)	*Origanum dictamnus* [Labiatae]
veronica (dried leaves)	speedwell (E.), véronique (Fr.)	*Veronica officinalis* [Scrophulariaceae]
juniper (berries)	genièvre (Fr.)	*Juniperus communis* [Cupressaceae]
nutmeg (seeds)	muscade (Fr.)	*Myristica fragrans* [Myristicaceae]
calamus (root)	sweet flag (E.), roseau (Fr.)	*Acorus calamus* [Araceae]
cinnamon (bark)	cannelle (Fr.)	*Cinnamomum zeylanicum* [Lauraceae]
peppermint (dried leaves)	menthe poivrée (Fr.)	*Mentha piperita* [Labiatae]

The manufacturing process

We will follow Duplais[25] and his recommendation for *Absinthe Suisse de Pontalier* (1885). Dried wormwood (2.5 kg), anise (5 kg), and fennel (5 kg) are macerated in 95 liters (L) of ethanol (85% by vol.) and steeped overnight in the pot of a double boiler. Water (45 L) is added and heat applied to the outer jacket to bring the decoction to a boil. The vapor is condensed in a coil with a room temperature water jacket. Heating the pot is continued until 95 L of distillate is collected. To 40 L of this distillate are added dried Roman wormwood (1 kg), hyssop (1 kg), and lemon balm (0.5 kg), which are extracted at a moderate temperature for some hours. After cooling, the liquor is siphoned off, filtered through a sieve of horse-hair, and reunited with the other 55 L of original distillate. A small amount of water is added to achieve a final alcohol concentration of 74% (by vol.); the yield is approximately 100 L. The process is simple and quick; no ageing of the product is required. According to Duplais[25] the distillers in Mont-pellier added coriander and angelica to the primary decoction; those of Lyon also added veronica to the secondary extraction step.

Three points are worth noting. First, by adding water to the first decoc-tion before heating, a type of "steam-distillation" was achieved wherein the amount of any constituent distilled over depends on both its vapor pressure and molecular weight. In this way the effect of a low vapor pressure for a particular compound may be counteracted to some extent by its high molecular weight relative to that of water.[26] Second, the distillation head of the industrial apparatus was simple and little attempt was made to restrict carry-over by aerosol entrainment. And third, the purpose of the secondary extraction at moderate temperature was twofold, to achieve a green coloration and to add additional flavor. Consequently, it is not a simple matter to predict the concentrations of some compounds in the final liqueur; in some cases it has been necessary to simulate the industrial process on a laboratory scale; see for example Arnold et al.[27]

Photographs and illustrative engravings from the turn of the century depict whole buildings for storing dried herbs, large (floor-to-ceiling) stills served by steam pipes, huge water-cooled condensers, and rail tankers of industrial alcohol to accommodate the massive production in some factories. Many of the stills were set up in tandem so that one could be engaged in distillation while its companion was being cleaned and recharged with herbs. On the other hand, there were also several small commercial producers and home-style operators.

According to a contemporary commentator on matters alcoholic almost every apothecary in every small French town had a liqueur resembling absinthe which was touted as good for everything that ails you.[28] Their products were made by mixing the various plant essences (essential oils made by steam distillation, which were commercially available) with industrial alcohol. The green coloration was achieved with chlorophyll from grass, parsley, nettles; virtually any available plant. Unfortunately, the smaller fabricators of absinthe were more prone to adulteration. Some unscrupulous dealers apparently achieved the *right* green tint by adding copper salts; on the bases of color and solubility in alcohol I would say that the particular compound used must have been normal cupric acetate, $Cu[CH_3COO]_2.H_2O$. According to *The Lancet*[17] a significant concentration of antimony was discovered in some batches of French absinthe. The journal thought that it was a misplaced attempt to add tartar emetic, antimony potassium tartrate, which at one time was inflicted on drunkards to "quiet their fury," and in this case putatively added as a built-in antidote. But a more likely candidate was antimony trichloride, $SbCl_3$. That salt is soluble in alcohol and not very soluble in water. It was probably used to achieve a better turbidity (louche) effect, upon dilution of the tainted absinthe with water.[23]

Finally, the quality of the grain alcohol used in absinthe manufacture varied, especially with respect to the content of fusel oils. All grain alcohol contains some fusel oil, which is a mixture of volatile alcohols other than ethanol, plus esters and aldehydes. The alcohols are derived from amino acids in the original plant material and are formed as by-products of yeast metabolism during fermentation. In relatively small concentrations they induce headache, fatigue, and nausea; and are said to contribute to *hang overs*. Acute intoxication with the alcohols of fusel oil leads to central nervous system depression and narcosis, hypotension, nausea, vomiting, and diarrhea. A typical fusel oil contains about 70% amyl alcohols [mostly 3-methyl-1-butanol and 2-methyl-1-butanol] and 20% isobutyl alcohol [2-methyl-1-propanol]; the LD_{50} values for a single oral dose to rats are 5.7, 4.0, and 2.5 g/kg respectively. There are much smaller amounts of 1-propanol and 2-phenethyl alcohol with LD_{50} values of 1.9 and 1.8 g/kg. The amount of fusel oil in the final distillation product depends not only upon the starting material but also on the still, the heat source, and the format employed by the distiller.

Methanol is in a special category and implies frank adulteration with wood alcohol. The LD_{50} for a single oral dose of methanol to rats is about 13 g/kg, not much different from ethanol. A dose of 80–200 g is fatal to man. However, chronic ingestion of relatively small concentrations of methanol causes headache, fatigue, nausea, and visual impairment. Acute intoxication leads to all of the above plus complete blindness, convulsions, respiratory collapse, and death. These damaging effects are due to the metabolic products formaldehyde and formic acid. The strong flavor of any absinthe could mask the taste of inferior alcohols.[13]

Notwithstanding the contribution of adulterants, Magnan and others found that wormwood extract was the primary culprit in connection with absinthism. In one paper essence of hyssop is mentioned in passing as being a precipitant of epilepsy in the experimental dog, but it was dismissed as being four or five times less active than wormwood essence.[29] It is probably safe to say that several of the constituents contributed by the other herbs are more or less toxic but the demonstrable effects of wormwood terpenes captivated subsequent investigators. In other words, the toxicity of the wormwood component was bad enough without need to look much further. Before discussing thujone and other compounds contributed by wormwood extracts, it is worth a brief look at the plant itself, and the long history of its association with man.

Wormwood and other *Artemisia* species

Artemisia absinthium is an herb with an almost woody base that reaches a height of no more than three feet. The root system is perennial and new shoots arise each spring. The flowers are yellowish-green, tiny, globular, but not visually outstanding. The leaves are highly indented and possess a silvery

ARTEMISIA.

Figure 4.2. *Artemisia* (the genus of Wormwood), woodcut, after Castore Durante, 1602, *Herbario Novo*, courtesy of Linda Hall Library, Kansas City, Missouri.

sheen. A compendium on cultivation, harvesting, natural pests and diseases of wormwood is available in Heeger.[30] The species is native to Europe and was more or less cultivated there from the Middle Ages until the second decade of the twentieth century (see Figure 4.2). But its association with man goes back to the earliest records.

Wormwood had religious and medicinal significance to the ancient Egyptians; earliest records appear in the Papyrus Ebers for which copies date from 1550 B.C. but include material from as early as 3550 B.C. The name of the genus derives from Greek mythology, i.e. *Artemis*, goddess of the moon, wild animals and hunting, *Apollo's* twin sister. She is identified with the Roman Goddess *Diana*, hence Shakespeare's "Diana's bud" in the potion of *Midsummer Night's Dream* is none other than wormwood:

Oberon.
But first I will release the fairy queen
Be as thou wast wont to be;
See as thou wast wont to see:
Diana's bud o'er Cupid's flower
Hath such force and blessed power.
Now, my Titania; wake you, my sweet Queen.

(Act IV, scene 1)

Wormwood is also mentioned in *Romeo and Juliet,* as a bitter extract applied to Lady Capulet's breast to wean Juliet, and in *Love's Labour Lost,* as a metaphor for Berowne's mocking tongue.

Seven biblical references to wormwood compare the plant's bitterness with sin and its consequenes. For example, Solomon, warning against unchastity, instructed:

For the lips of a strange woman drop as an honeycomb and her mouth is smoother than oil: But her end is *bitter as wormwood,* sharp as a two edged sword.

(Proverbs 5: 3–4).

The species name, *absinthium,* comes from the Greek *psintos* for delight. Pliny's *Historia Naturalis,* from the first century A.D., described *Artemisia* extracts as having long-standing utility against gastrointestinal worms, hence wormwood.[*] According to Pliny, "there is not a decoction of any herb of so great antiquity as it [wormwood] and none to have been used so long." It certainly works as an anthelminthic, terpenes from various *Artemisia* species are effective in stunning roundworms (nematodes) which are then expelled by normal peristaltic action of the intestine.[31]

About 65 A.D., Dioscorides finished his *De Materia Medica,* the first herbal and a most influential authority in pharmacy for the next 1,700 years. Both Pliny and Dioscorides recommended wormwood as an anthelminthic and an insect repellent. On the other hand, their lists of "virtues" were long and the remaining applications, including "it killeth any itch," "it saveth people who have taken hemlock," and "it provideth urine," remain unsubstantiated. Their books were set in type soon after the invention of printing in Europe, *circa* 1450. In the following century several derivative herbals appeared; others broke new ground.[32] They remarked upon the unique taste imparted to wine by an extract of wormwood and *Absinthites,* as Pliny called it, was embraced for medicinal value. Those decoctions were probably of relatively low toxicity.

However, after the publication of Hieronymus Brunschwig's two books *On Distillation* in the sixteenth century, the preparation of spirits and

---

[*] Parasitic worms (helminths) presently contribute to suffering, loss of productivity, and general poor health in many third world countries. Therapies based on terpene-containing plants have, to a large extent, been replaced by those using modern drugs (e.g. mebendazole, pyrantel pamoate and piperazine citrate) with similar action but with the advantage of negligible side effects, compared with terpenes such as camphor, thujone, and santonin, which will be discussed later.

concentrated essential oils from various plants including wormwood, became commonplace. Active, concentrated, and relatively stable medicaments based on plant compounds were feasible and were soon tested. Moreover, the potential for ingesting toxic doses was increased by orders of magnitude.

Several of the early herbals warned about the danger of taking *Artemisia absinthium* extracts in excess, but it was Johan Lindestolphe in 1708, with his *De Veneris* (On Poisons) who first cautioned that wormwood essence will lead to "great injury of the nervous system." He experimented upon himself and claimed an unusual sensitivity in that he could never taste the essence of wormwood without being "affected with headache and inflammation of the eyes." Lindestolphe and his follower, Christianus Stenzelius (his *Liber de Venenis*, 1739, was actually an edition of Lidestolphe's manuscript with additional commentary) also attested to narcotic and debilitating effects on many others. The warnings were unheeded and the "interesting" taste imparted to beverages by wormwood was frequently rediscovered in the next hundred years. Undoubtedly the induction of mood swings contributed to the popularity of these alcoholic drinks, but the putative medical value was always warmly received.

A more sophisticated recipe, involving wormwood and several additional herbs, was invented by a general practitioner in Switzerland, in the late eighteenth century. This was the forerunner of the absinthe that was popularized in France by troops who had been engaged in Algerian conflicts in the 1840's. The soldiers had sworn by absinthe or absinthe essence (as an additive to their wine) for its powers to combat and avoid fevers.[*] This form of "preventive medicine" quickly attracted adherents in Paris, and spread to the rest of France. The demand was met by eager producers in both Switzerland and France.

There are about 400 species of *Artemisia*. *A. maritima* was the species of wormwood used in Purl, a fortified ale, in Ireland and England up till the eighteenth century; it was reputed to be more intoxicating than regular beers or ales.[28] Purl refreshed Peter Simple in Shakespeare's *Merry Wives of Windsor*. *A. maritima* was also cultivated in the nineteenth century for the

---

[*]   For 2,000 years Chinese scholars extolled the virtues of *Artemisia annua* in the treatment of malaria. The efficiency of decoctions was confirmed in 1971; the active principle, qinghaosu, has been identified as a sesquiterpene lactone peroxide, and other derivatives have been prepared.[33] Malaria was common enough around the Mediterranean in the nineteenth century but *A. absinthium* is not a significant source of qinghaosu, and it is unlikely that absinthe was a febrifuge for the French North African troops, at least in connection with malaria.[34]

production of santonin, a powerful anthelminthic. The wormwood of biblical times is now thought to be *A. judaica*, and descendants still grow wild in Egypt and the Middle East. Some of the 200 species that flourish in parts of the United States include *A. dracunculus* (tarragon), *A. ludoviciana* (white sage), *A. filifolia* (sagebrush), *A. tridentata* (big sagebrush), *A. abrotanum* (southernwood), and *A. vulgaris* (mugwort).

Essential oils from *Artemisia* species are still used as fragrances, but their use as intentional flavor additives has practically ceased. In 1973 the Council of Europe[35] published a guideline for natural flavoring substances. *Artemisia absinthium* was listed within the category of spices or seasonings which can be added to foodstuffs in small quantities with a limitation on the active principle, in this case thujone, which must not exceed 10 parts per million in the final product. *Artemisia pontica, Tanacetum vulgare* (tansy) and *Salvia officinalis* (sage) have the same designation and the same restriction based on their thujone content. The concentration of thujone in absinthe was about 260 ppm.[*]

Essential oils, terpenes, and chemistry

Plants contain volatile, odoriferous oils, the so-called essential oils. For centuries the flowers, fruits, leaves and roots of selected plants (and to a lesser extent animal sources such as fish liver) have been processed for their essential oils. Oil of turpentine was certainly known to the Ancient Greeks more than 2,000 years ago. Crude or refined products of this sort are the mainstay of the flavor and fragrance industry; constituents of these essential oils are also used as starting materials for a variety of chemical syntheses.

The methods of extraction of essential oils from plants range from simple expression from the rind of fruit, through dissolution with fat solvents, to steam distillation. The latter is the most relevant to our present discussion. Plant material (usually dried and finely divided) is macerated in water, added to the pot of a still, and heated to boiling. The vapors pass through

---

[*] Notwithstanding the considerable literature on absinthe there is a paucity of published analyses on its most toxic component, thujone. My estimate is calculated from the following data. Each 100 liters of absinthe employed 2.5 kg of dried *A. absinthium* (1.5% oil, of which 67% is thujone) and 1 kg of dried *A. pontica* (0.34% oil, of which 25% is thujone). The contributions are 0.251 and 0.0085 g per liter respectively, making a total of 0.2595 g thujone per liter or 260 ppm. The two wormwoods also contributed about 90 ppm of thujyl alcohol, also a convulsant.

an overhead tube to a water-jacketed worm-shaped coil where they condense back to the liquid phase. This distillate usually separates into an essential oil above a water-rich bottom layer. Variations upon this process include passing live steam directly through stacked plant material or through a macerate in water. The principles of steam distillation[26] are applicable in all cases. As discussed earlier, a simple pot still allows some carry-over of less-volatile constituents due to entrainment of aerosol (droplets of the boiling liquid) in the moving vapor. Depending upon the nature of the essential oil and its application, the primary distillate may be refined by one or more subsequent procedures including fractional distillation, solvent extraction, or crystallization.

The constituents of essential oils fall roughly into three classes, in the following order of increasing prevalence. First, unbranched, long-chained aliphatic compounds, e.g. violet leaf aldehyde (**2,6 nonadien-1-al**)[*] and certain nitrogen- and sulfur-containing compounds, e.g. **allyl isothiocyanate** (from mustard oil), Figure 4.3. Second, aromatic compounds, i.e. derivatives of benzene, e.g. **anethole**, the major flavor constituent in anise. And third, the largest group, terpenes, which are variously hydrocarbons, alcohols, aldehydes, ketones, lactones or oxides, *not* aromatic in chemical character, and they may be solid or liquid at room temperature.[36, 37]

The main classification within the terpenes is as follows. Monoterpenes are C_{10} (i.e. 10 carbons from two C_5 **isoprene** units), sesquiterpenes are C_{15} (three isoprene units), diterpenes are C_{20} (four isoprene units), and so on. Within the monoterpenes there are acyclic, monocyclic, and bicyclic examples. The utility of the so-called isoprene rule was first formulated by Otto Wallach in 1887; for this and other advances he was awarded the Nobel Prize in Chemistry for 1910. Thus the head to tail association of two isoprene molecules gives rise to the acyclic compound **myrcene** whereas different associations give rise to isomers of myrcene or cyclic compounds such as **limonene**, Figure 4.3. The five-carbon fragments can be emphasized by bold bonds. The biosynthetic routes are more complicated, but many proceed via a C_{10} intermediate called **neryl pyrophosphate**, which gives at least an indication of the molecular architecture.

Otto Wallach's connection with art started in high school and his appreciation grew with his scientific development.[38] Wallach completed his doctoral work in record time under the direction of Professor Wöhler[**] and

---

[*]   Chemical structures are given for those compounds in bold face (see Figures 4.3, 4.4, 4.6, 4.7).

[**]   Friedrich Wöhler (1800–1882) synthesized urea with inorganic reagents in 1828. This was the beginning of organic chemistry.

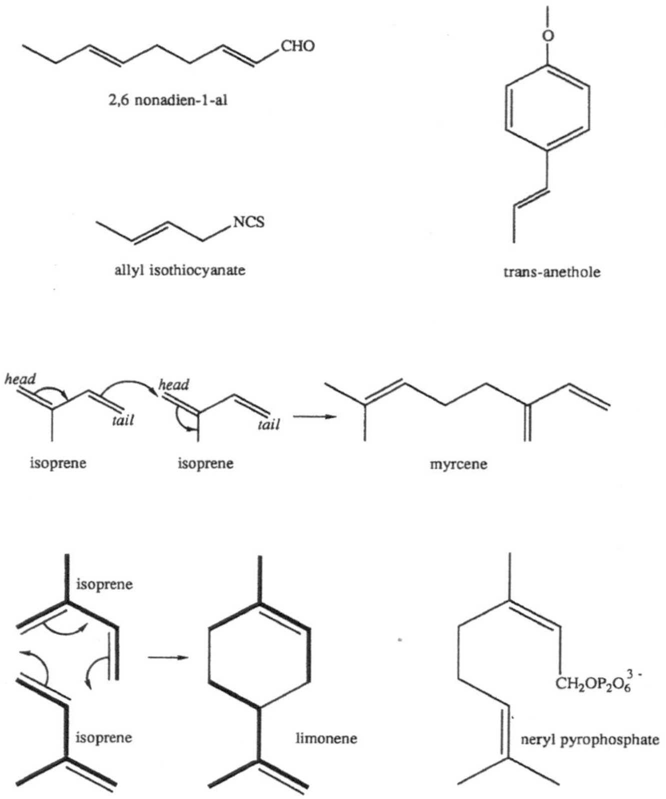

2,6 nonadien-1-al

allyl isothiocyanate

trans-anethole

isoprene   isoprene   myrcene

isoprene   isoprene   limonene   neryl pyrophosphate

$CH_2OP_2O_6^{3-}$

Figure 4.3. Examples of chemicals found in essential oils.

responded to the invitation of Professor Kekulé,[*] "It will not hurt you to come to Bonn. Here we lead a scientific artist-life." Otto Wallach later amassed a fine collection of watercolors in Göttingen, where he lived a bachelor in the old home of Friedrich Wöhler, within walking distance of the laboratory; he died there in 1931. How pleasant it is to contemplate the philosophies of Professors Kekulé, Wallach, and many of the other chemists of that time; their views had much in common with another contemporary, Vincent van Gogh.

---

[*] Friedrich Kekulé von Stradonitz (1829–1896) wavered between chemistry and architecture as a profession, and then with an artist's vision created structural organic chemistry through his conception of the ring structure of benzene.

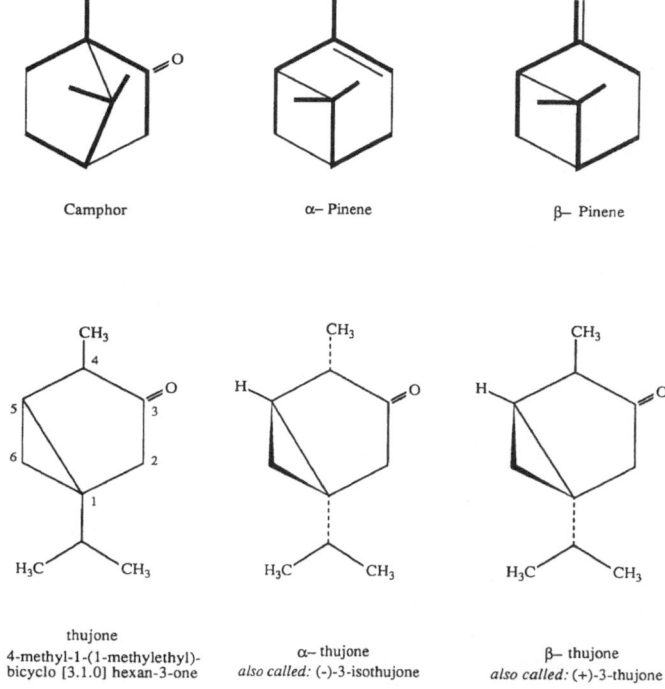

Figure 4.4. Camphor, pinene, and thujone.

The complications of bicyclic monoterpenes are exemplified by the travail that attended structure proof of **camphor** (Figure 4.4). After some 30 incorrect proposals by others, success fell to Julius Bredt,[39] in 1893. The key was to place the isopropyl group into the ring thus initiating a new type of bicyclic ring structure. The following year, Wagner proposed the correct structure for α-**pinene**. In 1899, Wagner and Slawinski were the first to suggest the proper structure for β-**pinene**. (Again, the bold bonds indicate the isoprene building blocks; they have no other significance and this artificial device will not appear in subsequent chemical structures.)

The correct formula for thujone was worked out by Friedrich Semmler,[40] of Greifswald, in 1900. Initially, Semmler (Figure 4.5) called the compound tanacetone because he worked with tansy oil, from the herb *Tanacetum vulgare.* Wallach and others called it thujone, for the oil of a tree *Thuja*

Figure 4.5. Protrait of Friedrich Semmler, after illustration in the report of Schimmel & Co. on essential oils, Leipzig, 1931 edition. Photography courtesy of R. Gelius.

occidentalis, a conifer in the arbovitae group. It soon became clear that thujone and tanacetone were the same compound presented in different mixtures of optical antipodes.

The structure of **thujone** is shown in Figure 4.4. The formal name is 4-methyl-1-(1-methylethyl)-bicyclo [3.1.0] hexan-3-one. Thujone has three asymmetric carbon atoms (1,4, & 5) but, because of restrictions imposed by the three-membered ring, only two isomers are found in nature. Considerable confusion in the literature resulted from three different systems of nomenclature, I have elected to use the simple, but less rigorous, α- and β- designations. The distinction resides in the orientation of the methyl group at C_4, in the β-isomer this bond (dashed) recedes from the viewer. Thuja and sage oils are richer in α-**thujone**; tansy and wormwood oils are richer in β-**thujone**. However, the isomers are interconvertable, the inversion about C_4 being promoted by acid, alkali, and the heat of steam distillation; the equilibrium is 33% α-thujone and 67% β-thujone (Figure 4.4). Rice and

Wilson,[41] tested the two isomers against mice and found remarkable differences in LD_{50}, minimum convulsive dose, and antinociceptive activity.

The louche effect and the ritual of presentation

In the fashionable method for mixing absinthe, a measured volume of the liqueur is placed in a special glass and then about five volumes of cold water are added slowly, over a cube of sugar, on a slotted absinthe spoon. The terpenes are freely soluble in the 74% v/v alcohol of the neat liqueur, but dilution with water brings them out of solution. The louche (turbid) effect, which follows as the terpenes are forced out of solution into a colloidal suspension, results in a yellow opalescence.[34] On the other hand, Toulouse-Lautrec's 'earthquake' stayed clear and green because he mixed absinthe with brandy, and the alcohol concentration remained sufficiently high to keep the terpenes in solution.

Sugar was added to old absinthe because the liqueur was quite bitter. The bitter principle, from wormwood, is absinthin, a complicated triterpene lactone $C_{30}H_{40}O_6$, whose complete structure was worked out comparatively recently by Czechoslovakian chemists (see Merck Index, 1989, item 7). Absinthin has a bitterness threshold of 1:70,000 and justifies the biblical analogy of "bitter as wormwood."

The ritual of dilution was part of the enthralling procedure that amused the absinthe drinker. The "Old Absinthe House" in New Orleans still maintains some of the old marble water fountains on the bar. Absinthe glasses, spoons, and water decanters are now collector's items – signs of a bygone day.

The chemicals of absinthe and related compounds

The importance of thujone in the toxicity of absinthe has already been mentioned. Some of the other terpenes encountered in the final product have not been adequately tested to be fully excused but the essential oils from the wormwoods, the contributors of significant amounts of thujone, are obviously the most culpable. The range of chemicals (mostly terpenes) that were encountered in absinthe can be gauged from Table 4, which is by no means exhaustive.

Table 4. Chemicals associated with the primary herbs of absinthe

Essential oil	Constituents [concentration]
wormwood[42]	thujone (67%); **thujyl alcohols and esters** (17%); sesquiterpenes (2%); limonene (1%); pinenes (1%); and 43 others in smaller amount
Roman wormwood[43,44]	**artemisia ketone** (47%); thujone (30%); and **1,8-cineole** (12%)
anise[45]	trans-anethole (92%); sesquiterpenes (2%); estragole (1%)
fennel[45]	trans-anethole (48%); **fenchone** (15%); thujene plus pinene (12%); limonene (10%); **estragole** (2%)
hyssop[36,46]	**pinocamphone** (45%); smaller amounts of pinenes, camphene, *p*-cymene
lemon balm[47]	**neral** (38%); **geranial** (31%); **caryophyllene oxide** (7%)

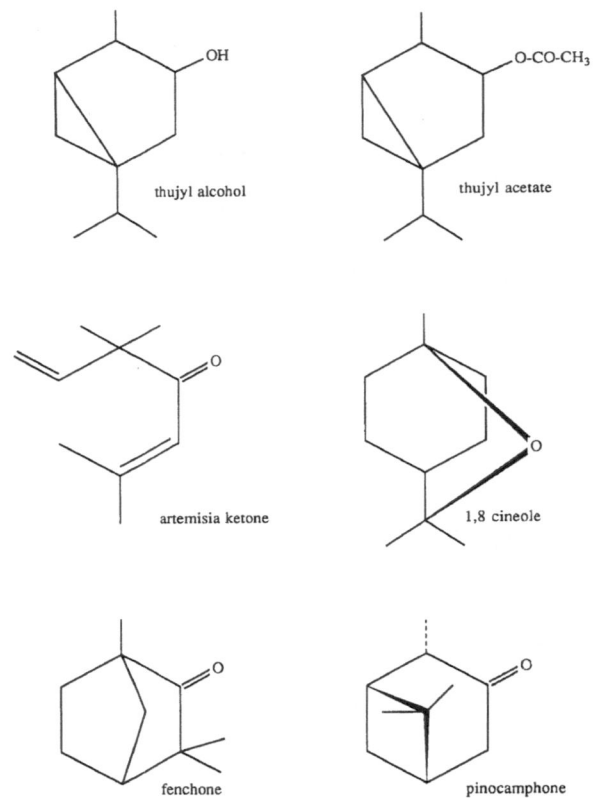

Figure 4.6. Chemical structures of selected terpenes identified in the text.

The content of thujyl alcohol (Figure 4.6) and its esters contributes to the toxicity of the wormwood component. Depending upon the location, Roman wormwood oil may contain as much as 24% 1,8-cineole (eucalyptol)[44] which is a dehydration product of **1,8-terpin**. Eucalyptol is in current use as a flavor for mouthwash and terpin hydrate has long been exploited as an expectorant. Terpin hydrate is listed in the U.S. and French Pharmacopoeia of 1890 and may have been a component in the cough drops with which Theo van Gogh had such a reaction in his final year (see chapters 3 and 5).

The secondary essences of absinthe also added a variety of chemicals:[45] Chinese star anise oil has 86% *trans*-anethole and 7% estragole (Figure 4.7); coriander 69% **linalool**, 6% limonene and 4% 1,8-cineole; angelica 39% limonene, 35% β-**phellandrene** and 9% α-pinene; and juniper essence is

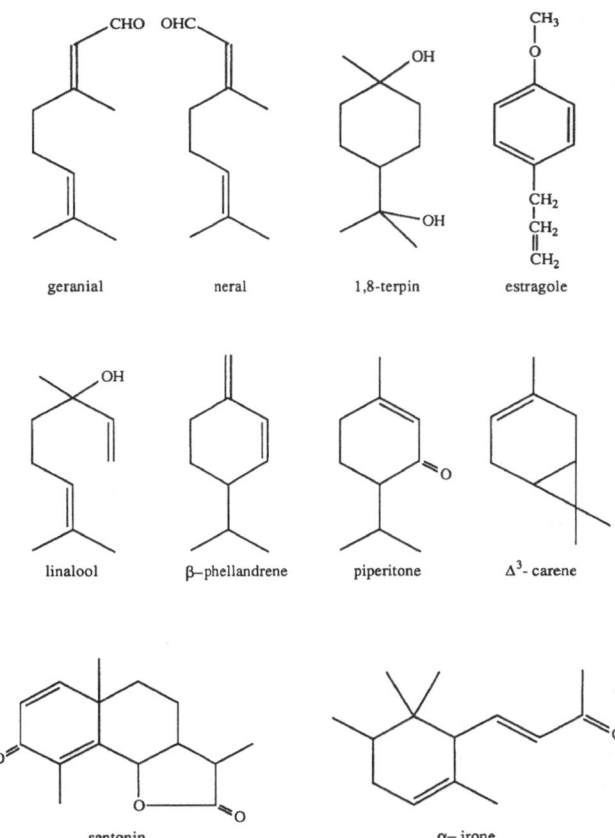

geranial   neral   1,8-terpin   estragole

linalool   β–phellandrene   piperitone   Δ³- carene

santonin   α– irone

Figure 4.7. Chemical structures of selected terpenes identified in the text.

71% α-pinene, 14% β-pinene, 3% limonene and 3% myrcene. These compounds contributed flavor and fragrance to various absinthes, according to the formulation.

The volatile oil of *Artemisia judaica* (biblical references) is about 1.6% (v/w) of the dried leaves and flowering heads and the major constituents are **piperitone** (40%), **camphor** (15%), and caryophyllene (6%) and borneol (4%). The presence of pipertone is a unique feature of this species of *Artemisia*. No thujone was found, but note that the pharmacodynamics of camphor and thujone are very similar; more about this below. The authors of this analysis[48] used material from a Cairo market and mentioned that the Egyptian plant is still commonly used in home remedies for its anthelminthic and other properties.

Artemisia maritima contains a substantial amount of **santonin**, a sesquiterpene lactone which was used as an anthelminthic for over 100 years. Several other *Artemisia* species produce santonin or related isomers[49] which will be discussed in further detail in chapter 8; on the other hand our own analyses[27] suggest that santonin was a very minor component of absinthe.

Turpentine and camphor

The composition of turpentine depends upon the grade and the source. Gum turpentine, the steam-volatile fraction of the gum that exudes from *Pinus palustris* (and other *Pinus* species) contains about 60% α-pinene, 30% β-pinene, and small amounts of monocyclic terpenes and terpene alcohols. Oil of turpentine (spirit of turpentine, essence of turpentine) is the fraction which distills between 154–170°C, and is mostly α- and β-pinene, camphene, and other terpenes including Δ^3 **carene** (Figure 4.7). It has a characteristic odor and taste and has been used as an antiseptic, anthelminthic, rubefacient and counterirritant. Pliny the Elder referred to turpentine's "virtues." Today, it is of much industrial value as a solvent, a vehicle for commercial paints, and a starting material for industrial syntheses (including camphor). Rectified oil of turpentine is made by treatment with sodium hydroxide and redistillation to remove unpleasant odor and taste. Oil of turpentine is also used as thinner or diluent of artists' oil paints and varnishes.

The ingestion of turpentine causes the urine to smell of violets. Cleopatra, queen of Egypt, was reputed to resort to this exotic device. In the case of accidental poisoning with oil of turpentine the odor is said to be pungent, persistent for weeks, and a diagnostic indicator.[50] The responsible compound is thought to be **irone**, derived by oxidation from a minor component (carotenoids). Ingested 1,8-cineole also imparts an odor of violets to the urine.

The therapeutic use of camphor in the Orient antedates history. It was not known to the ancient Greeks, but came into European use in medieval times. Castore Durante's *Herbario Nuovo*, 1585, has a plate correctly depicting *camphora* as a gummy exudate from a tree, thus ending much doubt and mystery in Europe about the origin of oriental camphor. Natural camphor is produced by steam distillation of wood chips from *Cinnamomum camphora*, which is native to Southeast Asia; today, it is also chemically synthesized from pinene. Pure camphor presents as a translucent mass with crystalline fracture. It has a slightly bitter and cooling taste. Oil of camphor is a by-product of the distillation process and contains 1,8-cineole (35%) plus significant amounts of pinene, camphor, phellandrene, dipentene, terpineol, safrole, eugenol, and sesquiterpenes.[51]

Kerosene

This is not a terpene, but is worth mentioning in connection with Dr. Peyron's final notation on Vincent van Gogh, at St. Rémy Asylum, wherein he mentioned the artist's attempts to eat his paints (mixed with turpentine) and also to drink petrol (probably kerosene) used to fill the lamps. Kerosene is one of several products produced by fractional distillation of crude petroleum oil. All of these fractions are primarily straight chain hydrocarbons of the alkane series. Kerosene is the fraction between nonane (C_9) and hexadecane (C_{16}), and also contains appreciable amounts of three different alkyl derivatives of benzene, naphthalene, and 1- and 2-methyl-5,6,7,8-tetrahydronaphthalene. It has a characteristic odor which, depending upon the preparation, is somewhat aromatic and not completely disagreeable.

Terpenes, epilepsy research, and convulsive therapy

The investigations of Magnan and others in the last half of the nineteenth century had shown that convulsions resembling epilepsy could be induced in man and experimental animals with toxic doses of absinthe. The essential oils were implicated, then essence of wormwood species in particular, and finally one chemical, thujone, bore the brunt of the criticism. Quantitatively speaking this is justified, but we should add a caveat about some of the other constituents of absinthe. Magnan[29] said that hyssop oil (which suggests pinocamphone) could precipitate convulsions if used in large enough amounts, and Wortis et al.[52] subsequently made passing reference to a similar toxicity of fennel oil (fenchone). To these can be added camphor, a minor component of absinthe and, from an entirely different source, carvone (dill and caraway oils). On the other hand pinene (turpentine) has scarcely any effect in this regard.[52] Turtschaninow and others found that santonin could evoke violent starts in experimental animals.[53] The cautious conclusion should be that any of the terpene ketones or lactones will probably have an adverse effect on the nervous system if taken in sufficient quantity.

Ingestion of excessive amounts of camphor in man leads to muscular symptoms but also hallucinations, flights of ideas, and mental confusion. The Czech physiologist Johannes Purkinje (1789–1869) took 2.4 grams of camphor before breakfast and described the results as follows:

> Movements seemed much facilitated; in walking my thighs rose far above the usual height, ... sensibility of the skin and muscle nerves seemed somewhat dulled. A storm of thought reached a degree to make loss of consciousness seem imminent, but consciousness was not actually lost because vomiting was induced ... condition continued for four hours, when a feeling of oppressive warmth spread over the whole body, and consciousness was lost ... When I awoke it took a long time to discover my own personality in relation to the time and place, for the whole morning and night formed a breach in the sequence of my life, and lay in dark uncertainty in my soul, which laboured to reconstruct my personality. Otherwise I felt well, and by no means fatigued, as in the case of other forms of inebriation.[53]

All of this type of experimentation involved acute doses. For example, various investigators determined the amounts of crude essential oils, or relatively pure terpenes, which would induce convulsions in an experimental animal. Cats, rabbits, dogs, and rats were variously favored for experimentation and different methods of administration were used; the more elegant trials standardized techniques, expressed results on an animal weight basis, and considered the animal's history. In the first half of this century thujone toxicity was a given, but related research focused on the development of an animal model for epilepsy.[53] Most of the physiological studies were devoted to thujone, camphor, and a man-made derivative, **3-bromocamphor**. Some representative results are summarized in Table 5.

Table 5. Terpenes causing epileptiform convulsions

Convulsant (formulation; method of administration)	Convulsion (type; concentration used; animal)
wormwood oil (5% v/v in 90% ethanol; intravenous)	minimal; 4.4 mg/kg; cat[54]
thujone (1% v/v in 6% w/v gum acacia; intravenous)	minimal; 3.4 mg/kg; rabbit[55]
thujone (2% v/v in propylene glycol; intravenous)	routine; 4 mg/kg; cat status epilepticus; 6 mg/kg; cat[56]
thujone (2% v/v in 6% gum acacia; intraperitoneal)	routine; 110 mg/kg; rat[57]
thujone (10% v/v in 95% ethanol; subcutaneous)	minimal; 50–100 mg/kg; mouse[41]
camphor (20% w/v in olive oil; intraperitoneal)	routine; 550 mg/kg; rat[57]
camphor (25% w/v in olive oil; intramuscular)	routine; 31–125 mg/kg; man[58]
3-bromocamphor (10% w/v in 95% ethanol; intravenous)	minimal; 5 mg/kg; cat[52]
3-bromocamphor (10% w/v in 95% ethanol; intravenous)	routine; 18 mg/kg; cat[56]

These results are much more sophisticated than the public demonstration which irked our friend Raoul Ponchon in his *Absinthe and the Guinea Pig,* but they do not put his skepticism to rest. The dangers are sufficiently apparent, but how do they relate to the absinthe drinker? To the best of my knowledge no studies have been reported on the chronic administration of absinthe, thujone, or camphor to experimental animals. The picture is incomplete. What we need to answer Raoul Ponchon is a long term (six months or a year) trial with daily administration of these chemicals, in amounts that would simulate the human consumption of several glasses of absinthe per day, and a careful evaluation of the animals at term. I am not aware of any studies along these lines; the drink was banned and the experimental interest moved to a more exciting development.

In the 1920's, the Hungarian neuropathologist Lászlo von Meduna made fundamental observations on brain tissues of patients with epilepsy as compared with schizophrenia. These basic investigations, and the concept that the two conditions were incompatible, prompted his exploration of induced seizures in the treatment of schizophrenics. At first these convulsions were provoked by administering camphor and then later he introduced metrazole as the convulsive agent.[59] A retrospective on more than 3,000 cases appeared in 1939 and indicated great benefit for certain forms of schizophrenia.[60]

Pharmacologic convulsive therapy started with camphor, chemical cousin to thujone, evolved through the more soluble **metrazole** (pentylenetetrazole), and was optimized with **indoklon** (hexafluorodiethyl ether) administered by inhalation. The latter turned out to induce convulsions at 30 ppm in the aspired air and was in vogue in the late 1950's. Although electroconvulsive therapy (ECT) has replaced these chemical approaches, it is worth noting that the beneficial effect of such therapy is brought about by the convulsion itself rather than the flow of current or the chemical administered.[58] Wormwood and thujone have come full circle from the virtues of ancient anthelminthics and insect repellents, through the gloom of absinthism, to the positive contributions to modern medicine.

Figure 4.8. Vincent van Gogh: *Still Life, Drawing Board with Onions, Raspail's Book, Absinthe Bottle etc.* 1889, St. Rémy, oil on canvas, 50 × 64 cm (19.7 × 16.2″), Collection: Rijksmuseum Kröller-Müller, Otterlo, The Netherlands

Vincent van Gogh and the individual response

Some years ago, while perusing the letters of Vincent van Gogh, I was intrigued by the chemical connection between absinthe and some other terpenes in his life. These well-documented exposures involved Vincent's use of massive amounts of camphor to combat insomnia, an attempt to drink essence of turpentine (pinene), and references to his nibbling at oil colors (mixed with turpentine). The possibility of an interaction became more compelling when I read Sollmann[61] on thujone and camphor, wherein he remarked that the convulsions induced in experimental animals are antagonized by bromide, while the threshold is lowered by nicotine. (Van Gogh's crises were ameliorated by taking bromides and decreasing smoking, while institutionalized in Arles.) Accordingly I suggested that van Gogh had developed an affinity (pica)* for terpenes, the documented examples being thujone, camphor, and pinene. This would help to explain some of the strangest of van Gogh's acts during his last two years – his attempts to eat his paints and to drink turpentine and kerosene – which were previously regarded as absurdities and unrelated.[64]

The response to any drug depends upon a variety of factors not least of which the nutritional status of the subject. Accordingly it is intriguing to note that a 1912 report indicated that camphor is metabolized to **3-hydroxycamphor** (campherol) and secreted as **3-hydroxycamphor glucuronide**, a conjugate with glucuronic acid as shown in Figure 4.9.

An *increased toxicity of camphor* (and related compounds) was noted during fasting, and a compromise in glucuronic acid formation was suggested as the mechanism.[61] Infections and underlying illness also play critical roles in determining the individual's response to drugs. We shall see in the next chapter that certain inherited diseases confer unusual sensitivity in this respect, and that both alcohol and terpenes can precipitate medical crises.

---

* Pica comes from the Latin word for "magpie," a bird who carries away odd objects. It refers to compulsive eating of non-nutritive substances such as dirt (geophagia), ice (pagophagia), flaking paint, plaster, or clay (trichophagia), and has been ascribed to various disorders including malnutrition.[62] Carlander[63] mentions one hundred cases in Sweden which were all successfully treated with therapeutic doses of iron. The pica symptoms invariably disappeared in about a fortnight. Some of the patients also had anemia, which took longer to rectify. These observations suggest a more immediate connection between pica and sideropenia rather than anemia.

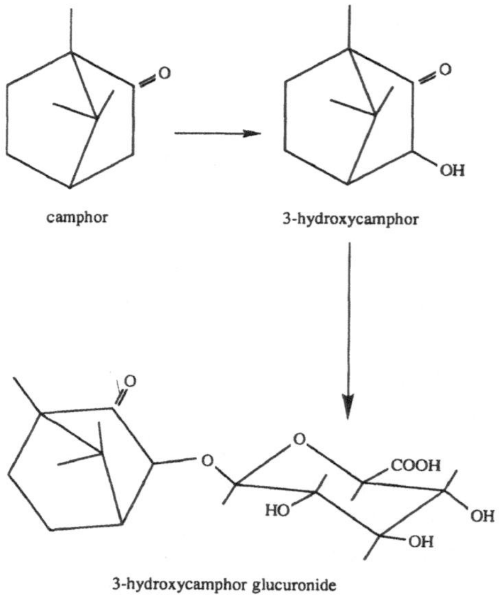

134

Figure 4.9. Proposed metabolic pathway for camphor.

References and notes

1. Schmidt H. 1915. l'Absinthe. l'Aliénation mentale et la criminalité. Rapport fait au nom de la commission d'hygiène publique de la chambre des députés. *Annales d'Hygiene Publique et de Médicine Légale* 4th series, **23**: 121–133.
2. Adams B. 1980. Picasso's absinth[e] glasses: six drinks to the end of an era. *Artforum* **18**: 30–33.
3. Motet A. 1859. *Considérations générales sur l'alcoolisme et plus particuliérement des effets toxiques produits sur l'homme.* Medical thesis: University of Paris.
4. Marcé [L]. 1864. Sur l'action toxique de l'essence d'absinthe. *Comptes rendus herbdomadaires des Séances de l'Académie des Sciences (Paris)* **58**: 628–629.
5. Magnan [V]. 1869. Epilepsie alcoolique; action spéciale de l'absinthe: épilepsie absinthique. *Comptes rendu des seances et memoires de la société de biologie (Paris)* 4th series, **5**: 156–161.
6. Magnan [V]. 1874. On the comparative action of alcohol and absinthe. *The Lancet* vol 2 for 1874: 410–412.
7. Magnan [V] & Fillassier A. 1912. Alcoholism and degeneracy. *International Eugenics Conference (London).* pp 367–379.
8. Amory R. 1868. Experiments and observations on absinth[e] and absinthism. *The Boston Medical and Surgical Journal* **78**: 68–71 & 83–85.
9. Ponchon R. 1925. *La Muse au Cabaret.* Paris: Bibliothèque-Charpentier, Fasquelle Éditeurs.
10. Goodwin D. 1988. *Alcohol and the Writer.* Kansas City & New York: Andrews & McMeel.

11. Flower D & Maas H. 1967. *The Letters of Ernest Dowson.* p 35, (letter to Arthur Moore, February 15, 1889). Cranbury, New Jersey: Associated University Presses Inc.

12. Wolff A. 1881. *Le Figaro (Paris).* April 10, as quoted by Moffett CS. 1986. *The New Painting: Impressionism 1874–1886.* San Francisco: Fine Arts Museums of San Francisco.

13. Brasher CWJ. 1930. Absinthe and absinthe drinking in England. *The Lancet* Vol 1 for 1930: 944–946.

14. Hulsker J. 1990. *Vincent and Theo van Gogh: A Dual Biography.* pp 401–404. Ann Arbor: Fuller Publications.

15. Hulsker J. *op. cit.* p 322.

16. Conrad B.III. 1988. *Absinthe, History in a Bottle.* San Francisco: Chronicle Books.

17. Anonymous. 1873. Absinthism. *The Lancet* Vol 1 for 1873: 22.

18. Anonymous. 1898. The consumption of alcohol in France. *The Lancet* Vol 2 for 1898: 1603.

19. Anonymous. 1906. Alcohol in France. *The Lancet* Vol 2 for 1906: 1531.

20. Anonymous. 1912. The consumption of alcohol in France. *The Lancet* Vol 2 for 1912: 794.

21. Anonymous. 1903. The absinthe evil. *The Lancet* Vol 2 for 1903: 620.

22. Delahaye M-C. 1983. *L'Absinthe Historie de la Fee Verte.* Paris: Berger-Levrault .

23. Anonymous. 1868. Absinthe. *American Journal of Pharmacy* **40**: 356–360.

24. Walker EE. 1906. The effects of absinthe. *Medical Record* **70**: 568–572.

25. Duplais P. 1855. *Traité des Liqueurs et de la Distillation des Alcools ou le Liquoriste et le Distillateur Modernes.* Versailles: Chez l'Auteur.

26. Maron SH & Prutton CF. 1958. *Principles of Physical Chemistry.* pp 168–170. New York: The Macmillan Company.

27. Arnold WN, Dalton TP, Loftus LS & Conan PA. 1991. A search for santonin in *Artemisia pontica,* the other wormwood of old absinthe. *Journal of Chemical Education* **68**: 27–28.

28. Saintsbury G. 1920. *Notes on a Cellar-Book.* London: Macmillan and Co. Ltd.

29. Magnan [V]. 1890. Des principaux signes cliniques de l'absinthisme. *Bulletin de la Société de Médecine publique et d'hygiene professionelle (Paris)* **13**: 349–363.

30. Heeger EF. 1956. *Handbuch des Arznei- und Gewürz-pflanzenbaues Drogengewinnung.* pp 268–274. Berlin: Deutscher Brauernverlag.

31. Baldwin E. 1943. An in vitro method for the chemotherapeutic investigation of anthelminthic potency. *Parasitology* **35**: 89–111.

32. Anderson FJ. 1977. *An Illustrated History of the Herbals.* New York: Columbia University Press.

33. Brossi A, Venugopalan B, Dominguez Gerpe L, Yeh HJC, Flippen-Anderson JL, Buchs P, Luo XD, Milhous W, & Peters W. 1988. Arteether, a new antimalarial drug: synthesis and antimalarial properties. *Journal of Medicinal Chemistry* **31**: 645–650.

34. Arnold WN. 1989. Absinthe. *Scientific American* **260**: 112–117.

35. Council of Europe. 1973. *Natural Flavouring Substances, their Sources, and added Artificial Flavouring Substances.* Strasbourg: Maisonneuve.

36. Simonsen JL. 1949–1957. *The Terpenes.* vol I–V. 2nd ed. rev. London & New York: The Syndics of the Cambridge University Press.

37. Pinder AR. 1960. *The Chemistry of the Terpenes.* New York: John Wiley & Sons Inc.

38. Ruzicka L. 1932. The life and work of Otto Wallach. *Journal of the Chemical Society* Part 1 for 1932: 1582–1597.

39. Bredt J. 1893. Über die Constitution des Camphers und einiger seiner Derivate. *Berichte der Deutschen Chemischen Gesellschaft* **26**: 3047–3057.

40. Semmler FW. 1900. Über Tanaceton und seine Derivate. *Berichte der Deutschen Chemischen Gesellschaft* **33**: 275–277.

41. Rice KC & Wilson RS. 1976. (–)-3-Isothujone, a small nonnitrogenous molecule with antinociceptive activity in mice. *Journal of Medicinal Chemistry* **19**: 1054–1057.

42. Srinivas SR. 1986. *Atlas of Essential Oils*. New York: Anadams. Note: Analyses of oils from wild wormwood, growing at different elevations, have also been reported:- Chialva F, Liddle PAP & Doglia G. 1983. Chemotaxonomy of wormwood, *Artemisia absinthium*. 1. Composition of the essential oils of several chemotypes. *Lebensmittel-Untersuchung und -Forschung* **176**: 363–366.

43. Hurabielle M, Tillequin F & Paris M. 1977. Étude chimique de l'huile essentielle d' *Artemisia pontica*. *Planta Medica Zeitschrift für Arzneipflanzenforschung* **31**: 97–102.

44. Chialva F & Liddle PAP. 1981. Sur la composition de l'huile essentielle de *Artemisia pontica* cultivé en Piémont. *Rivista Italiana E.P.P.O.S.* **62**: 350–352.

45. Formàcek V & Kubeczka KH. 1982. *Essential Oils Analysis by Capillary Gas Chromatography and Carbon-13 NMR Spectroscopy*. Chichester: John Wiley & Sons.

46. Hilal SH, El-Alfy TS & El-Sherei MM. 1980. Investigation of the volatile oil of *Hyssopus officinalis*. *Egyptian Journal of Phamaceutical Science* **19**: 177–184. Note: Pinocamphone (43%), isopinocamphone (31%) and small concentrations of 30 other terpenes were found in the oil from North American, cultivated hyssop:- Lawrence BM. 1984. Hyssop oil. *Perfumer and Flavorist* **9**: 38.

47. Enjalbert F, Bessiere JM, Pellecuer J, Privat G & Doncet G. 1983. Analysis of the essential oil of balm. *Fitoterapia* **54**: 59–65. Note: Other analyses have been summarized:- Lawrence BM. 1989. Melissa oil. *Perfumer and Flavorist* **14**: 75–76.

48. Karawya MS, Hifnawy MS, & El-Hawary SS. 1977. Volatile oil of *Artemisia judaica*. *VII International Congress of Essential Oils*, Kyoto, Japan, Item # 106, p 359. Note: Differences due to plant location have also been reported:- Fleisher Z & Fleisher A. 1990. The essential oils of *Artemisia judaica* from the Sinai and Negev Deserts. Aromatic plants of the Holy Land and the Sinai, Part II. *Journal of Essential Oil Research* **2**: 271–273.

49. Geissman TA & Irwin MA. 1970. Chemical contributions to taxonomy and phylogeny in the genus *Artemisia*. *Pure & Applied Chemistry* **21**: 167–180.

50. Graphel FG. 1901. Turpentine poisoning. *British Medical Journal* vol 1 for 1901: 340.

51. Anonymous. 1889. Notes on essential oils from Messrs. Schimmel & Sons annual report – "Camphor oil." *American Journal of Pharmacy* **61**: 313–314.

52. Wortis SB, Coombs HC & Pike FH. 1931. Monobrominated camphor. A standard convulsant. *Archives of Neurology and Psychiatry* **26**: 156–161.

53. Muskens LJJ. 1928. *Epilepsy*. New York: William Wood & Company.

54. Pike FH, Elsberg CA, McCulloch WS & Rizzolo A. 1929. Some observations on experimentally produced convulsions: the localization of the motor mechanisms from which the typical clonic movements of epilepsy arise. *American Journal of Psychiatry* **9**: 259–283.

55. Keith HM. 1935. Experimentally produced convulsions. Effect on thujone convulsions of insulin and of variations in water content of brain. *Archives of Neurology and Psychiatry* **33**: 353–359.

56. Opper L. 1939. Pathologic picture of thujone and monobromated camphor convulsions: comparison with pathologic picture of human epilepsy. *Archives of Neurology and Psychiatry* **41**: 460–470.

57. Sampson WL & Fernandez L. 1939. Experimental convulsions in the rat. *Journal of Pharmacology and Experimental Therapeutics* **65**: 275–280.

58. Kalinowsky LB & Hoch PH. 1961. *Somatic Treatments in Psychiatry. Pharmacotherapy; Convulsive, Insulin, Surgical, other Methods.* New York & London: Grune & Stratton.

59. von Meduna LJ. 1937. *Die Konvulsiotherapie der Schizophrenie.* Halle: C. Marhold.

60. von Meduna L & Friedman E. 1939. The convulsive-irritative therapy of the psychoses. A survey of more than three thousand cases. *Journal of the American Medical Association* **112**: 501–509.

61. Sollmann T. 1948. *A Manual of Pharmacology and its Applications to Therapeutics and Toxicology.* 7th ed. Philadelphia & London: WB Saunders Company.

62. Anonymous. 1959. Aetiology of pica. *The Lancet* Vol 2 for 1959: 281.

63. Carlander O. 1959. Aetiology of pica. *The Lancet* Vol 2 for 1959: 569.

64. Arnold WN. 1988. Vincent van Gogh and the thujone connection. *Journal of the American Medical Association* **260**: 3042–3044.

Chapter 5
The Diagnosis

Figure 5.1. Vincent van Gogh: *Self-Portrait with Japanese Print*, 1887, Paris, oil on canvas, 44 × 35 cm (17.3 × 13.8″), Dr. h.c. Emile Dreyfus-Stiftung, Kunstmuseum Basel

> *Our neurosis ... (is) ... a fatal inheritance.*
> Vincent to Theo, letter 481, from Arles,
> May 4, 1888.

Van Gogh's attending physicians in the south of France were terse in their hospital entries about his illness.[1] Felix Rey, a young intern, still in training at Arles and yet to receive his medical degree, took care of Vincent after his initial crises. Dr. Peyron, onetime military ophthalmologist and now in a second career, gave asylum at St. Rémy. They have received mixed evaluations.

At one extreme, the present-day proponents of "atypical epilepsy" for Vincent have deemed Rey to be brilliant and insightful. Their circular argument goes as follows: Rey embraced "epilepsy" without evidence of a full-fledged case; the commentators believe temporal lobe epilepsy (described many years later) is an attractive possibility; therefore they say, Rey was ahead of his time.[2] At the other limit, Peyron has been judged by some as naive and trained in the wrong specialty. Other commentators are more moderate and most agree that Rey and Peyron did their best to protect and rehabilitate the artist during those demanding two years.

Tralbaut[1] felt that these physicians were sympathetic to van Gogh's suffering but were not particularly interested in taking a complete medical history. He suggested that Rey, and then Peyron, were overly influenced by the police reports in Arles and by the patient's own preconceived notions of a family history of epilepsy, which was apparently gleaned from his mother. If so, then the circle was indeed completed when Vincent wrote to Theo, "As far as I can make out, the doctor here [Dr. Peyron] is inclined to consider what I have had [was] some sort of epileptic attack" (letter 591).

If indeed Drs. Rey, Urpar, and Peyron were convinced that Vincent van Gogh had some sort of epilepsy then why wasn't he treated for it? Admittedly the available therapy at that time was meager, but Vincent was not even treated symptomatically at St. Rémy, and no advice along those lines was passed on to Paris when Vincent departed. Contrast Vincent's case with that of Dostoevsky* (1821–1881) who wrote, on June 17, 1863, "I go to Paris and Berlin ... only for consultation of specialists (Trousseau in Paris, Romberg in Berlin) for my epilepsy."[3] It is curious that Vincent, a quarter of a century later, was not referred to an epilepsy specialist at Montpellier or Paris!

---

*   Vincent van Gogh was inspired to paint the fever ward at the hospital in Arles (the canvas was actually completed later in St. Rémy) after reading an article about Dostoevsky and his book, *Souvenirs de la maison des morts,* (letters W15, 535).

Dr. Louis Rivet had received a reasonable exposure to psychiatry in his formative years in Paris (see chapter 7 for further discussion on Vincent's doctors). It is assumed that Theo kept him more or less informed of Vincent's condition, but it is again curious that Rivet apparently played no role during the last few months. Instead, Paul Gachet became the doctor of merit.

Dr. Gachet had speciality training in mental ailments but he knew Vincent just for the last two months, in Auvers-sur-Oise. There is no evidence that Paul Gachet had any direct communication with either Rey or Peyron, or that Vincent carried any medical documentation when he came north. This seems to be remarkably cavalier behavior in retrospect. However, Gachet had conversed earlier, and at some length, with Theo van Gogh who relayed the news to Vincent: "I met Dr. Gachet … When I told him how your crises came about he said to me that he didn't believe it had anything to do with madness and that, *if it was what he thought, he could guarantee your recovery,* (my italics) but that it was necessary for him to see you and to speak with you in order to be able to make a more definite statement" (letter T31). Would that we knew whether Dr. Gachet was simply fashioning an optimistic prognosis for Theo or whether his own written notes were more conservative! I assume that Paul Gachet kept a medical journal but it has not been found.

It seems inconceivable that Dr. Gachet kept no records. His son and daughter maintained the residence in Auvers-sur-Oise after the doctor's death in 1909, and they were renowned for the care, nay, reverence, with which they preserved items of their father's clothing (e.g. the white cap in the van Gogh portrait), medical instruments, and memorabilia. They had no children and were survived by distant relatives. Rumor has it that somewhere along the way all of Dr. Gachet's records were intentionally destroyed "to protect the privacy of his patients." His views survive only in the form of interesting anecdotes, with poor documentation of time or place.

On the other hand, posthumous hypotheses and speculations about the nature of van Gogh's sickness have been profuse.[1,4] They range from romantic explorations, through loosely collected signs and symptoms and vague declarations of syndromes, to concise scientific statements. Some have claimed that van Gogh had a disease as unique as his art, which does not advance the field one iota. Still others seek to protect Vincent from the stigma, which they imagine would attend certain diagnoses and thereby detract from the value of his art. Accordingly, they apparently feel obliged to obstruct any attempt to increase the knowledge base, and they will not be dignified by any further attention. The more reasonable of past suggestions will be evaluated in the next chapter. Here, I shall repeat the hypothesis of Loftus[*] and Arnold,[6] and discuss the medical and biochemical features of acute intermittent porphyria (AIP), the disease entity which I continue to believe most closely fits Vincent's case.

Vincent's illness

The letters[7] remain the best source of unembellished descriptions of his medical problems, albeit presented in lay terms. It is axiomatic that any reasonable working hypothesis must address all of the information; this includes family history and the artist's life style as well as the underlying illness. Hemphill[8] was probably the first to apply this principle in the van Gogh context. The interaction between congenital disease and exacerbating factors is central to my argument. The importance and influences of malnutrition and alcoholic beverages were brought out in chapters 2, 3, and 4. A brief recapitulation follows.

Vincent's ailment was characterized by episodes of acute mental derangement and disability which were separated by intervals of lucidity and creativity. Moreover, attending physicians, family, friends, and the artist himself were all surprised and encouraged by the rapidity of the recoveries after each crisis.[1,4,7] His serious illness developed late in the the third decade, as evidenced by his concern with "the possibility that [my] family might take steps to deprive me of the management of my affairs and put me under guardianship" (letter 204), (and see also chapter 2, wherein serious problems are indicated at age 27). His underlying complaint was characterized by frequent gastrointestinal problems (see, for example, letters 448, 530, B4), and at least one bout of constipation that required medical intervention.[9] The condition caused fits with hallucinations, both auditory and visual, (see, for example, letters 592, W11) and evoked partial seizures.[10] Periods of incapacitating depression and physical discomfort were severe and grave enough to provoke self-mutilation[11] and eventual suicide[7] (see also chapters 9 and 10). Some of his bouts of sickness may have been

---

[*] Dr. Loretta S. Loftus suggested AIP in December 1988, after reading my paper, *Vincent van Gogh and the thujone connection.*[5] As a practicing internist she was immediately impressed by the frequency of references to neuropsychiatric and gastrointestinal complaints by Vincent, the involvement with alcohol, and the intermittent nature of his psychosis. Together we searched the letters for everything on Vincent's illness. The artist's symptomatic agreement with the hallmarks of acute intermittent porphyria, the recognition of precipitant factors in his life, and the van Gogh family history of mental illness, all supported the case for AIP.

associated with fever (letter 206) and sexual impotence (letter 506). His ailment was exacerbated by overwork (letter 173), malnutrition and fasting (letters 440, 571), environmental exposure (letter B15), excessive ingestion of alcoholic beverages (for example, letter 581), especially absinthe (letter A16), and a proclivity for camphor and other terpenes.[5] The symptoms were palliated during institutionalization with better diet, abstinence from alcohol (letters 595, 599), and administration of bromide therapy (letter 574). In spite of their severity he did not experience any permanent, functional disability after any attack.[1,4,7] There was a family history of mental illness.[1,4,7]

Notwithstanding the restraints of retrospective analysis on a definitive diagnosis, the gamut of symptoms and responses is consistent with acute intermittent porphyria, a hereditary metabolic disease that provides a unifying hypothesis.[6] The first case was reported by Stokvis[12] (see Figure 5.2), just one year before van Gogh's death, but AIP did not enter the practicing diagnostician's armamentarium until decades later.[13]

Figure 5.2. Portrait of Barend Josef Stokvis, about 1902.

Acute intermittent porphyria

This is one member of a class of metabolic abnormalities, the porphyrias, which are characterized by the excessive production of porphyrins, or related compounds. Individuals who suffer from these diseases are prone to excrete elevated concentrations of these same compounds in their urine and feces. The abnormal excretion per se is of no intrinsic medical import but it is a reflection of elevated concentrations circulating within the body, and therein lies the potential for cutaneous photosensitivity (due to porphyrins), neurological abnormalities (due to porphyrin precursors), or both. In the case of AIP, all of the symptoms are neurological and the specific, overly-produced compounds are δ-aminolevulinic acid (an amino acid) and porphobilinogen (a pyrrole). Waldenström[14] noted that AIP should be called a "pyrollia," but this more rigorous nomenclature has not been adopted. These are intermediates in the pathway to porphyrins, which in turn are used in the biosynthesis of the heme of hemoglobin, and other heme-containing proteins.

"Acute" refers to the rapid onset, and abrupt cessation, of expressed symptoms. (The underlying cause of AIP is present from birth, so in that sense it is chronic.) "Intermittent" refers to the periodicity, which is typical, and emphasizes the distinct periods of normalcy, which usually intercede between the episodes of expressed symptoms. The expression of symptoms very much depends upon life-style, nutrition, and exposure to alcohol and drugs. The following modern cases of AIP are well documented, both clinically and biochemically, and have been selected as bearing resemblances to the illnesses of Vincent, Theo, or Wil. Each medical history has been shortened and paraphrased; the original publications should be consulted for fuller detail.

Analogies with Vincent van Gogh's illness

Case 1 [late 1970's]

Compare with: "I am often terribly melancholy, irritable ... I am terribly sensitive, physically as well as morally, the nervousness being developed during those miserable years which drained my health" (letter 212). Vincent to Theo, July 6, 1882.

A 31-year-old man had been hospitalized eight times in the preceding six years with disorientation, confusion, depression, auditory hallucinations, bizarre and persecutory delusions, and peculiar behavior. He had occasional abdominal pain but no unusually colored urine. This man was only discovered to have AIP during a survey of institutionalized psychiatric patients. AIP was confirmed chemically and enzymatically. The clinical examination at that time showed marked depression, mood-incongruent auditory hallucinations, but excellent insight.[15]

Case 2 [1977]

Compare with: "However, the unbearable hallucinations have ceased and are now getting reduced to a simple nightmare in consequence of taking bromide of potassium I think" (letter 574). Vincent to Theo, January 28, 1889.

A 38-year-old man with known AIP, whose mother also had the disease, suffered an unwitnessed episode of loss of consciousness, in February 1977, which was associated with abdominal pain and pink-discoloration of his urine. In September of the same year he had major seizures. On admission to hospital he had orthostatic hypotension and tachycardia, plus colicky abdominal pain. During the next four days, a high carbohydrate diet was associated with decreases in porphobilinogen (PBG) (40%) and δ-aminolevulinic acid (ALA) (26%) excretion, but on the fifth day he had a tonic-clonic seizure. He later described a premonitory feeling of "closing in" and "impending doom" with a sensation of nausea. Hematin therapy produced a prompt decrease in the excretion of PBG and ALA, but one or two daily seizures continued and seemed to be exacerbated by the initial drug therapy. On the seventeenth day, generalized tonic-clonic and complex partial seizures continued and the patient's speech and behavior deteriorated. Large doses of potassium bromide were substituted for previous drugs, after which seizures ceased and behavior improved. The dose was adjusted to maintain an appropriate level of bromide in the circulating blood. After six months without seizures, the patient stopped taking bromides and then seizures recurred two weeks later. Bromide therapy was restarted.[16]

Analogies with Theo van Gogh's illness

Case 3 [1949–50]

Compare with: "[Theo had] serious nervous afflictions, so bad that he could not move ... still stiff, as after a fall." Andries Bonger, December 31, 1886. "Theo's cough was worse than it was when

I left him two years ago." Vincent to Wil,
June 1890, (letter W22).
"For a week he [Theo] was unable to
urinate." Camille Pissarro, October 18,
1890.

A man of 29 years first consulted his family physician about muscular weakness in September 1949. He also complained of frequent sore throats. A tonsillectomy was performed in October 1949, but one month later the patient returned with complaints of nausea, vomiting, and acute abdominal pain. Ingestion of greasy foods aggravated intestinal upsets. The gall bladder was removed. About the beginning of 1950 the patient again developed nausea, vomiting, acute abdominal pain and weakness in the right arm. His wife noticed that his urine was often brown. By the time he was admitted to hospital paralysis had progressed throughout the whole body. Urological examination revealed numerous purplish-red spots throughout the bladder mucosa. (This finding, by cystoscopy, was supposedly due to extreme bladder distension.) A specimen of urine taken by catheter was clear and pale yellow but turned brownish in transit to the laboratory and a diagnosis of AIP was then suggested. This was confirmed by chemical analysis and phenobarbitol (a known precipitant of AIP crises) therapy was stopped. However, the paralysis progressed rapidly and included loss of sphincter control. The patient died on the fourth day after admission. Several members of the immediate family had suffered similar conditions and had died of undiagnosed diseases.[17]

Case 4 [1949]

Compare with: "The drops [of medicine] of that Dr. [Van der Maaten] ... helped to stun me during the night and prevented me from coughing, but they gave me hallucinations and nightmares ... I was literally crazy." Theo to Wil, September, 1890.

A woman of 23, with a history of abdominal pain and vomiting of one week's duration, was admitted to hospital. (The year before she had a similar occurrence and was treated for "kidney trouble" with bed rest.) On the basis of a tentative diagnosis of gallstones, a laparotomy was performed, but the biliary tract proved normal and no abdominal abnormality was found. Her urine gave a strong reaction for porphobilinogen and AIP was suggested. The patient had a persistent postoperative cough which developed into pneumonia. She had difficulty coughing and her sclerae assumed a plum-colored tinge. She then made a partial clinical recovery and got up, but later

148

complained of numbness in both thighs at night, and a week later could hardly raise herself in bed. After two more days she developed tachycardia and the next week abdominal pain returned. Nine days later she lost her voice, and never regained it. She had great difficulty in swallowing and died after a week with generalized flaccid paralysis. The mother had died of a "stroke" at 49. Two cousins on the mother's side died at 23 and 21 of "acute ascending myelitis" and "anxiety neurosis plus pneumonia."[18]

Analogy with Wil van Gogh's illness

Case 5 [late 1970's]

Compare with chapter 4: Wil showed signs of "insanity" at the age of 41 and was committed to an asylum in The Hague. At the end of 1902 she was accommodated at Veldwijk, a mental home in Ermelo, run by the church. She died there in 1941 having spent the latter half of her 79 years in a mental asylum.

An elderly man had been institutionalized for at least 40 years with aggressive, inappropriate behavior and with periods of withdrawal and depression. At the time AIP was diagnosed (due to a survey), and confirmed chemically and enzymatically, the medical chart noted that the patient maintained an embryonic position, displayed chewing movements of his mouth, moaned, was nonresponsive, avoided contact with others, was ambulatory but resisted care. His mother had been hospitalized for a psychotic depression.[15]

Comments on the case studies

AIP is still a difficult diagnosis, the intervals reported between initial symptoms and the correct diagnosis range from several months to years,[19] and as much as 40 years (case 5). Because of their psychoses, many patients are institutionalized (cases 1 and 5) and many of these are never properly diagnosed. Because abdominal pains are frequent, AIP patients are often referred to the surgical ward, and come under the knife unnecessarily (cases 3 and 4). Many otherwise useful drugs will actually precipitate crises (cases 2 and 3).

Not all patients exhibit reddish or dark urine (case 1); freshly voided urine is often of normal color but darkens upon standing (case 3). Direct light accelerates this process. Urine which has aged internally due to bladder dysfunction may already be discolored when released with a catheter. However, few patients with AIP call attention to their own dark or reddened urine (case 3) and the color is sometimes mistaken for urinary bleeding.

Convulsions are not uncommon (case 2), but occur in only about 10% of all AIP cases.[14] The triad of abdominal pain, peripheral neuropathy, and psychiatric disturbance is the most typical presentation, but the clinical manifestations of AIP are extremely varied, and individual illnesses may mimic several other diseases.

Clinical manifestations of AIP

Symptoms rarely occur before puberty; the peak decade for onset of symptoms is from age 20 to 29 (somewhat higher for males than females), as shown in Figure 5.3, but the disease sometimes remains latent throughout a lifetime.[14, 20]

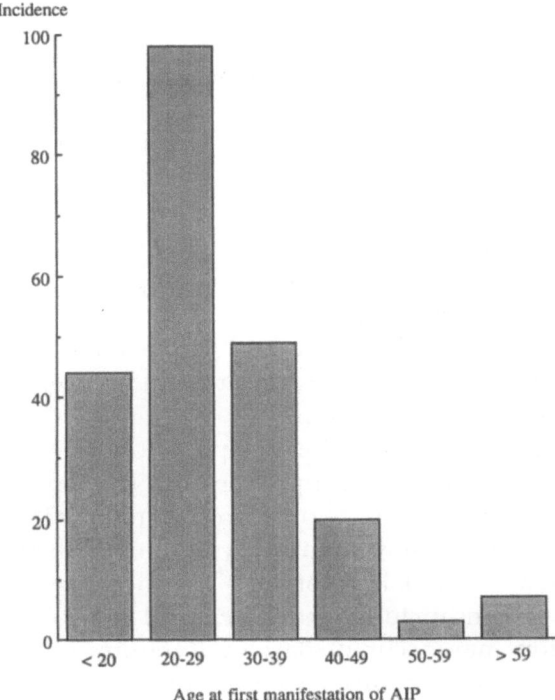

Figure 5.3. Age of onset of acute intermittent porphyria, adapted from Waldenström.[14]

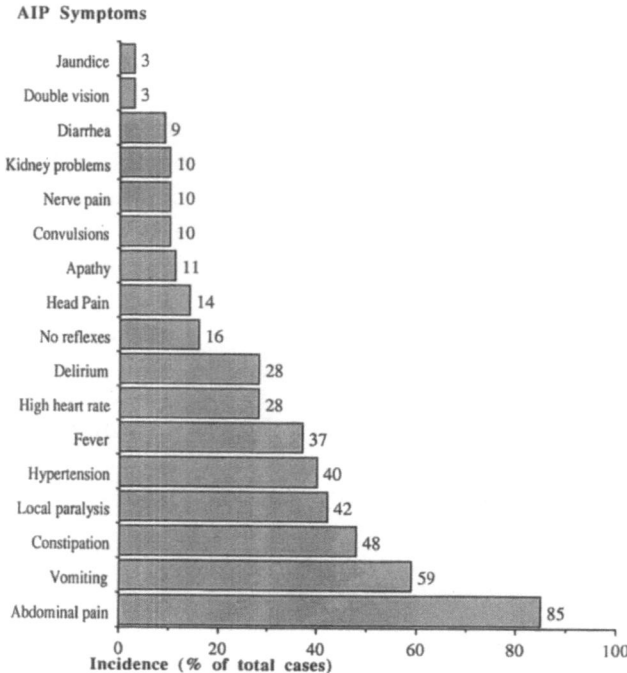

AIP Symptoms

Figure 5.4. Incidence of symptoms in acute intermittent porphyria, after Waldenström.[14]

Tabulations (Figure 5.4) of the most common hallmarks emphasize abdominal pain and other gastrointestinal complaints, symptoms referable to the peripheral and central nervous systems, and signs of autonomin neuropathy including tachycardia and hypertension.[14,20] Porphyria-induced hypertension can cause early-onset renal failure.[21] Bladder dysfun - tion may result in urinary retention.[20,21]

Effects on optic nerves or the occipital lobes leading to blindness (amaurosis),[22] drooping of both upper eyelids from paralysis of the third cranial nerve (bilateral ptosis),[23] dilation of one pupil,[23] and involuntary rapid, rhythmic eyeball movements (nystagmus)[23] have all been documented for AIP cases. Sexual impotence[20] has occasionally been reported. Premonitory symptoms include restlessness and irritability; attacks develop rapidly; resolution may occur in days or sometimes weeks, in an unpredictable fashion. Seizures do not always attend severe crises, but when they do antiseizure drugs, with the notable exception of bromides,[16,24] may adversely affect the outcome.[20,24]

The unpredictable nature of the disease with respect to both onset of crises and outcome makes an acute attack of AIP particularly treacherous. It can be one of the most terrifying experiences imaginable. Patients can become almost completely paralyzed in severe cases. They are unable to breathe, swallow or communicate properly, yet remain conscious for some time, all the while suffering pain, being aware of their plight, and wondering if it will ever end. The most common cause of death from AIP is respiratory paralysis.

Exacerbating factors

Most importantly, the expression of neurological and other symptoms depends upon life-style and exposure to precipitating factors. Early examples of AIP were revealed as a response to new drugs; initially the hypnotic Sulfonal, later barbiturates, and subsequently many other drugs, alcohol, and sundry organic compounds.[24] Some steroid metabolites precipitate attacks and endogenous changes may account for some crises at puberty, the earlier onset with females, and relationships with the menstrual cycle and oral contraceptives. Other exacerbating factors include infections and malnutrition.[20] Low-carbohydrate and low-protein diets are especially detrimental,[25] and fasting can precipitate an attack of porphyria.[26] A study in Scotland[27] indicated an association between smoking and the induction of repeated attacks in patients already diagnosed with AIP.

Vincent van Gogh and AIP

All of the hallmarks of Vincent's illness can be accommodated within this overview of AIP. The most important and well documented are the gastrointestinal complaints, neurological disturbances, age of onset, jagged time course, and the exacerbations caused by inadequate nutrition and absinthe abuse. Other aspects such as sore throats, eye problems, fevers, a bout of aphasia in the Arles hospital, and impotence, have other possible causes but are all compatible with an underlying illness of AIP. Van Gogh's smoking habit may have contributed to recurrent attacks.

Vincent's urinary tract infection in The Hague may have precipitated an AIP crisis leading to the "complication" and extended hospitalization at that time. It is possible that his urinary retention was a consequence of an AIP attack. The suggested primary diagnosis of gonorrhea is even a little suspect since there was no mention of a purulent discharge.

Dark or reddish urine is not mentioned in the published letters. However, it may have escaped attention. Vincent's accommodations were often primitive; for example, the Yellow House in Arles had no toilet and he was forced to use the facilities at the hotel next door (letter 480). We assume he relieved himself in the field while painting. Furthermore, even freshly voided urine during crises of AIP is often normal color, the pigmentation comes with retention in the bladder or upon ageing.

Arnold[5] suggested that van Gogh's fondness for absinthe developed into a pica for terpenes, the documented examples being thujone, camphor, and pinene. It is worth noting that 1,8-cineole, (the chemical structure for 1,8-cineole [eucalyptol] is given in Figure 4.6) a constituent of crude camphor and wormwood oils, is a proven precipitating agent for AIP.[28] Van Gogh used reckless doses of camphor oil against insomnia (letter 570) and absinthe contained a variety of essential oils including wormwood. The combination of overexposure to camphor, absinthe abuse, and fasting or malnutrion would be injurious for anyone, but devastating for someone with AIP.

Theo van Gogh and AIP

Tralbaut[1] and others noticed some similarity between the illnesses of Vincent and Theo. There were numerous exchanges between the brothers concerning their "nervous" problems. It is not clear whether Theo's serious illness at age 19 was related to the expression of AIP-like symptoms but certainly by age 29 he had "serious nervous afflictions, so bad that he could not move" according to his future brother-in-law (see chapter 3). Within two months of Vincent's suicide Theo suffered further leg pains and hallucinations (partly in response to an unspecified medicament for his cough), became very irritable and occasionally violent, muttered with difficulty in mixed languages, experienced urine retention, and was totally unconscious with a barely detected pulse before he died (aged 34). Leg pains, mental illness, and paralysis would all support a diagnosis of AIP, and the violent reaction to a new drug and renal failure would be compatible with AIP.

The time course of Theo's illness was obviously different from that experienced by Vincent, but nevertheless both can be accommodated within the spectrum of established AIP cases. Goldberg and Rimington[29] summarized 50 cases and found that a few patients merely felt occasional abdominal pain while at the other extreme others had explosive, fatal attacks which lasted ten days to ten weeks. Two cases exhibited a stepladder pattern, wherein multiple attacks occurred over a period of months, each more severe than the preceding, until the final paralytic and fatal attack.

Statistics indicate that early onset is the most sinister prognostic indicator; patients who express symptoms in the second decade have a mortality rate of over 60%.[29] The majority experience recurring attacks with interceding periods of months or years, with restitution of physical and mental function. Many patients recover quickly and completely. Some are severely crippled with muscle wasting and weakness. The overall mortality rate is about 24%.[29]

Wil van Gogh and the other siblings

Vincent's youngest sister, Wil, spent the latter half of her 79 years in an asylum for psychiatric cases. She may also have suffered from AIP, although the lack of further documentation makes her case much more speculative. The youngest brother, Cor, died at 33 in South Africa from an accident while feverish; it may have been suicide (see chapter 10). Again, the medical history is scant. His other sisters, Elizabeth and Anna, lived 77 and 75 years respectively, without any indication of medical crises.

The biochemical lesion in AIP

Almost any cell in the human body can engage in heme synthesis, this biosynthetic pathway is not only vital for the synthesis of the heme of hemoglobin but also for the cytochromes involved in so many aspects of metabolism. However, the major sites are the bone marrow and the liver, and the problems which arise have been characterized as either erythropoietic or hepatic respectively. We are concerned here with an example of the latter.

An abbreviated representation of the biosynthetic pathway for heme is shown in Figure 5.5. The key enzyme for our purposes is porphobilinogen deaminase which directs and catalyzes the joining together, head to tail, of 4 molecules of porphobilinogen to form hydroxymethylbilane. This intermediate is subject to non-enzymatic tetrapyrrole ring closure to form uroporphyrinogen I, (the depicted pathway has been simplified) but in the presence of another enzyme, uroporphyrinogen III cosynthetase, the major product is uroporphyrinogen III. The latter has an interesting asymmetry (note the relative positions of the acetate [A] and propionate [P] substitutions) brought about by a specific rearrangement under the direction of the cosynthetase.[30] The process is further complicated by a compartmentalization within cells; thus the early and final steps occur within sub-cellular organelles (mitochondria) whereas the intermediate events proceed in the soluble part of the cell (cytosol). Traffic of molecules in and out of the

154

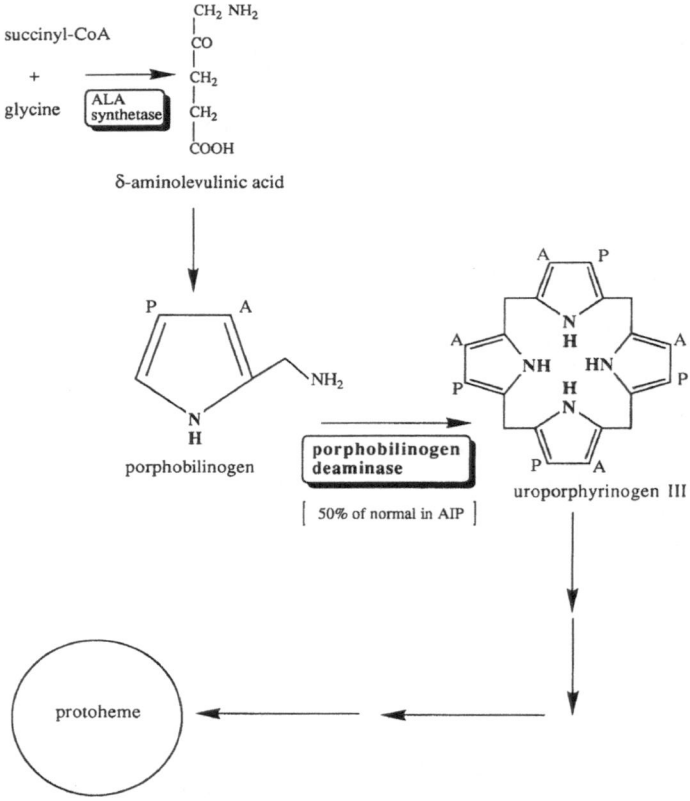

Figure 5.5. The heme pathway. Key enzymes associated with acute intermittent porphyria are indicated.

mitochondrion is part of the process and one aspect of orchestration for the overall pathway.

Each of the chemical reactions (arrows) indicated in Figure 5.5 is catalyzed by a specific protein (an enzyme). The enzymes themselves are manufactured by a separate cellular machinery. When all of the heme pathway enzymes are in place the flow of carbon starts with a small amino acid (glycine), progresses through porphobilinogen (a pyrrole), gives rise to uroporphyrin and other porphyrin intermediates, and finally arrives at heme (protoheme plus an iron atom). The dynamics are of special interest because they explain the way in which cells can control the amount of heme manufactured, the problems associated with upset of any one particular step, the role of interfering substances (the exacerbating factors of disease), and the disease modifiers (compounds useful in therapy). In a healthy individual, without chemical or physical insult, the synthetic machinery performs like a well disciplined orchestra.

The first enzyme, δ-aminolevulinic acid (ALA) synthetase, catalyzes the reaction which limits the overall rate of heme synthesis. This enzyme is present in lower amount than any of the subsequent enzymes. In part this is due to rapid degradation of the enzyme within the cell and accordingly it must be continually synthesized. Therein lies a major control feature, the regulation of ALA synthetase. Heme, the end product of the pathway, causes both a repression[31] (lower rate of synthesis of the enzyme) and an inhibition[32] (lower catalytic rate of the existing enzyme) of ALA synthetase. This phenomenon is called feed-back. Also, the synthesis of this key enzyme can be partially blocked by high glucose intake,[33] thus helping to explain the ameliorating effect of a high carbohydrate diet on AIP attacks, and the adverse effect of malnutrition and fasting. On the other hand some steroids will induce the synthesis of the enzyme as much as 40–50 fold.[34]

The individual with AIP has about one half the normal amount of porphobilinogen deaminase, the third enzyme in the heme pathway.[35] This is the basic, inherited defect. Some of the earliest data which demonstrate the deficiency and the consequence are summarized in Table 6.

Table 6. Liver metabolism

Subjects	δ-aminolevulinic acid synthesis (%)*	porphobilinogen conversion (%)*
Controls (non-porphyric patients)	100 ± 13 (n = 12)	100 ± 4 (n = 9)
Acute intermittent porphyria		
patient #1	680	40
patient #2	680	42
patient #3 (in crisis)	1167	11
Porphyria cutanea tarda**		
patient #4	112	118
patient #5	95	120
patient #6	85	118

* Rates are expressed relative to controls. Data adapted from Strand et al.[35]
** This is a different form of porphyria which is associated with abnormal skin fragility and sensitivity to sunlight. Note that rates for ALA synthesis and PBG conversion are normal.

Note that the rate of PBG metabolism is decreased to about 40%. This is due to a deficiency in porphobilinogen (PBG) deaminase. In patients with

AIP the level of ALA synthetase tends to be elevated, this is reflected in the increased (sevenfold) metabolic rates. Patient #3, showed even more dramatic changes while undergoing a crisis. Subsequent investigations of liver, red blood cells and a variety of other cell types revealed the consistent finding that PBG deaminase activity was about 50% of normal in clinically expressed AIP, or latent AIP, subjects. The method based on red blood cells is the most convenient and least invasive.[36]

Because PBG deaminase is not rate limiting to the overall pathway, 50% of normal may still be sufficient for unstressed AIP patients to supply their heme needs. This explains the lack of symptoms for latent AIP patients and the intervening periods of normalcy for patients who have expressed episodes of sickness. The exacerbation factors exert their effect by increasing the synthesis of ALA synthetase and, in combination with the underlying deficiency in PBG deaminase, lead to an accumulation of ALA and PBG. An increase in the first enzyme, which is normally rate-limiting, and a partial block in the third enzyme, causes the two intermediates ALA and PBG to build up and then "spill" into the urine. Under these circumstances excretion in the urine of these two metabolites is significantly increased. Freshly voided urine is normal color but with time PBG polymerizes to form pigments which impart the dark appearance to aged specimens.

While the overproduction of ALA and PBG is clearly a prerequisite for neurological problems the mode of action is not completely understood at this time; at least three working hypotheses have been offered. First, ALA or PBG (or a yet to be identified derivative) may be toxic to the nervous system. Increasing amounts accompany acute attacks and, even though increased excretion is sometimes seen in latent cases, neurological dysfunction per se never occurs when porphyrin precursor excretion is normal.[37] ALA is structurally similar to γ-aminobutyric acid, a neurotransmitter and ALA may interfere with γ-aminobutyric acid-receptors on nerve cells.[38] An important criticism here is that neither ALA nor PBG appear to cross the blood brain barrier.[39]

A second suggestion arises from investigations of chemically induced hepatic heme depletion in the rat wherein it was found that liver tryptophan 2,3-dioxygenase (a heme-containing enzyme) activity is lowered.[40] This impairment spares tryptophan degradation, increases the concentration of tryptophan served to the central nervous system, and increases serotonin turnover. In the experimental system, exogenous heme reversed the effect whereas tryptophan potentiated it. The authors suggest that the serotonin mechanism may play a role in the neurological disturbances associated with AIP.[40]

Thirdly, if the deranged pathway for heme also operates in the nervous system, there may be a shrinkage of the heme pool locally and this may be the cause of neurological problems.[41] Regardless of the mechanism, the two forms of useful therapy, namely high carbohydrate feeding and hematin infusion, both act by suppressing the synthesis of ALA synthetase.[41]

Many drugs are metabolized in the liver by a system which involves a subcellular organelle (endoplasmic reticulum) and cytochrome P_{450}. This protein is the major consumer of heme in the normal, healthy liver and its induction by drugs and xenobiotics further increases consumption to the point of becoming a drain on the heme pool.[42] Alcoholics also exhibit a proliferation of endoplasmic reticulum[43] in their liver cells and part of the ethanol ingested is metabolized therein via a cytochrome P_{450} mediated system.* In this manner, alcohol and other drugs which lower the heme pool are thought to exert their influence on the overall pathway by derepressing (hence increasing) the synthesis of ALA synthetase. In the AIP subject, with half the normal amount of PBG deaminase, the insult may be too great and toxic levels of ALA and PBG may develop. One system that was developed expressly to test potential porphyrinogenic drugs depends upon measuring the induction of ALA synthetase in fetal chicken liver cells that have been exposed to test compounds.[44]

Precipitant drugs

The following are selected examples of drugs that may precipitate attacks of acute porphyria: all varieties of barbiturates; anticonvulsants such as the hydantoins and succinimides; anxieolytics such as chlordiazepoxide and meprobamate; oral contraceptives; steroids; several non-barbiturate hypnotics such as sulfomethane, trional, paraldehyde, and glutethimide; some antimicrobials including sulfonamides, griseofulvin, and erythromycin; antidiabetics such as tolbutamide; antihypertensives including methyldopa; various ergot preparations; several anesthetics including lidocaine (local), cocaine (topical), pentothal (intravenous), and halothane (inhalation); ethanol; and sundry examples from almost every other drug category.[24]

---

* In fact there is a competition for this system between alcohol, barbiturates, and some other drugs. The decreased rate of degradation of barbiturates explains their longer (and sometimes lethal) action if taken together with alcohol. The proliferation of the endoplasmic reticulum of alcoholics explains their increased tolerance of barbiturates because they have the machinery for increased degradation, provided they are not drinking ethanol at the same time.

Some unusual examples of precipitants

Pentylenetetrazole (Metrazole), which was introduced in chapter 4 in connection with pharmacologic convulsive therapy, caused an acute attack in a patient with AIP.[24] Eucalyptol (1,8-cineole), a constituent of *Artemisia pontica* oil, and crude camphor oil, was shown by Bickers et al.[28] to be the culprit in an unusual case involving a patient with AIP who drank a mouthwash containing this flavorful terpene. Mustajoki and Koskelo[45] reported on a fatal case of AIP wherein the precipitant factor was aspidiin, the active principle of *Aspidium* (male fern), employed as an anthelmintic.

Thujone, camphor, and pinene

Dr. Herbert L. Bonkovsky, University of Massachusetts Medical Center, and I tested these compounds against chicken embryo liver cells. In the presence of desferrioxamine, an iron chelator which inhibits heme synthesis and thereby mimics the block associated with acute porphyria, the terpenes enhanced porphyrin accumulation 5–20 fold. They also induced synthesis of the rate-controlling enzyme for the pathway, δ-aminolevulinic acid synthetase. Thus there is a significant porphyrogenic potential for thujone, camphor, and pinene as well as the previously described eucalyptol.[28] Vincent van Gogh exposed himself to all of these compounds.

Drugs and treatments thought to be safe for AIP patients

Specific treatments include high carbohydrate diets and infusions of hematin. Abdominal and other pains are treated with aspirin, morphine, other opium constituents and related compounds, and meperidine.[20,24] Episodes of vomiting can be relieved with chlorpromazine. Safe drugs for hypertension and tachycardia include guanethidine and propranolol. The treatment of neurosis, psychosis, and seizures is achieved with chlorpromazine, diazepam and bromides. Neostigmine is used for constipation. (In van Gogh's day the only available items from this group would be high carbohydrate diet, opium, and bromides.) Other drugs which are thought to be safe for patients with AIP include atropine, chloral hydrate, digitalis compounds, lithium, penicillins, quinine, and tetracyclines.

Genetics

Barker and Estes,[46] in 1912, were the first to note that AIP runs in families. The extensive studies of Waldenström[47] in Sweden firmly established the inherited nature of the disease. The disease follows an autosomal dominant pattern of inheritance; if one parent is a carrier then on the average 50% of the children will bear the defective gene. However, the penetrance is variable, with only a fraction of the carriers usually expressing the disease.[48] The enzymic lesion is porphobilinogen deaminase[39] and the gene for the enzyme is localized[49] to the distal long arm of chromosome 11. Mutation within this gene leads to a defective protein product, one that has no porphobilinogen deaminase activity. This aberrant product would be altered in one or more amino acids and may not survive long enough within cells to be detected by reaction with antibodies to the normal protein. Such seems to be the case in 80% of AIP patients thus far examined.[50] Most of the remainder exhibit a defective protein that cross-reacts with appropriate antibodies, and is present in about 60% the amount of normal enzyme. A very small group apparently overproduces a different aberrant protein or (more likely) makes an altered protein that has a longer life by resisting cellular degradation. Whether this heterogeneity accounts for differences among patients in the susceptibility to exacerbation factors, and the clinical presentation, remains to be elucidated.

Children who are AIP gene carriers remain clinically latent. Within adults with the disease there are two populations. The majority of subjects do not develop clinical expression. Within those who do express symptoms there are patients who also have a deficiency in hepatic steroid 5 α-reductase.[51] There is evidence in man for genetic control of responsiveness to steroid hormones within different tissues.[52] How a major defect in steroid 5 α-reductase is acquired or related to AIP is not clear, but is not due to abnormal porphyrin metabolism per se. A close correlation between elevated sex hormone-binding globulin and clinical expression of AIP has also been demonstrated.[53] It has been suggested that certain drugs may precipitate AIP attacks in part because they act through the endocrine system.[20]

A rare form of AIP is associated with the usual 50% deficiency of PBG deaminase in liver cells but normal amounts in bone marrow and circulating red blood cells.[54] One Finnish family was discovered with this defect.[55] Eight AIP patients from four families in northern Holland have been similarly documented; they are a part of a large kindred traced to a common ancestral pair in the early nineteenth century.[56]

Vincent's parents

As far as we know Vincent's mother led a seemingly healthy life and died at 88. However, according to the register at Arles "He (Vincent) told us that his mother's sister was epileptic, and that there were many cases in this (the mother's side of the) family."[10] His father, the Reverend Theodorus van Gogh, died at 63; his studies for the church had been interrupted by serious illness; he was judged by at least one commentator not to have been in very good health most of his life.[1] The Reverend Theodorus was said to have died from a stroke and, because hypertension is present in over half of AIP patients,[57] this underlying disease would be one of many possibilities compatible with the cause of death. Two of the father's family, Uncle Cent and Uncle Hein, suffered from ill health (undocumented), took early retirement, and died at 68 and 63 respectively. Of Vincent's parents the father may be the more likely (obligate) carrier of AIP, but this is little more than an educated guess. He led a careful and balanced life in his "post in the wilderness"[58] and may have avoided the precipitating factors that affected three of his six children. There may have been some other contribution from Vincent's mother which would explain the unusually high penetrance.

Summary

Loftus and Arnold[6] suggested that Vincent van Gogh suffered from acute intermittent porphyria, exacerbated by malnutrition and absinthe abuse. This provides a reasonable working hypothesis on the nature of his illness. All of the documented symptoms can be accommodated, and the family history provides additional support. The case for AIP in brother Theo is also compelling and, in some ways, easier to justify. Evidence for sister Wil is scant except that she exhibited aberrant behavior which caused her admission and long tenure in an asylum. There are of course many alternative explanations, but surveys conducted since then indicate a disturbing incidence of AIP in mental institutions. Brother Cor's possible suicide and its underlying causes remain a mystery (but see chapter 10 for more discussion).

Paradoxically, the diagnosis of AIP for Vincent van Gogh can be challenged because so many siblings expressed symptoms. As an autosomal dominant condition we expect, on the average, three of six children to carry the defective gene from the affected parent. On the other hand the expression within carriers of the same family has rarely been 100%. Perhaps the other parent contributed something that led to increased expression. This would explain the unusually high degree of penetrance.

The concept of sickness or something less than full health being compat-

Figure 5.6. Vincent van Gogh: *A Corner of the Asylum Garden*, 1889, St. Rémy, oil on canvas, 73.5 × 92 cm (28.9 × 36.2″), Museum Folkwang, Essen

ible with (letter 607) or even encouraging (letter 570) creative work is a recurring theme in Vincent's correspondence. While contemplating leaving the asylum at St. Rémy (see Figure 5.6) and a return to the North, he wrote, "As for eating a lot, I do – but if I were my doctor, I'd forbid it. I don't see any advantage for myself" (letter 607). If he suffered from acute intermittent porphyria his own advice was the worst possible.

162

References and notes

1. Tralbaut ME. 1981. *Vincent van Gogh*. New York: The Alpine Fine Arts Collection, Ltd.
2. For this point of view see Gastaut H. 1956. La maladie de Vincent van Gogh envisagée à la lumière des conceptions nouvelles sur l'épilepsie psychomotrice. *Annales Medico-Psychologiques* vol 1 for 1956: 196–238; wherein Dr. Aussoliel, a friend of Dr. Rey, was deemed to be a local expert on masked epilepsy.
3. Voskuil PHA. 1983. The epilepsy of Fyodor Mikhailovitch Dostoevsky (1821–1881). *Epilepsia* **24**: 658–667.
4. Lubin AJ. 1987. *Stranger on the Earth: A Psychological Biography of Vincent van Gogh*. New York: Henry Holt & Co.
5. Arnold WN. 1988. Vincent van Gogh and the thujone connection. *Journal of the American Medical Association* **260**: 3042–3044.
6. Loftus LS & Arnold WN. 1991. Vincent van Gogh's illness: acute intermittent porphyria? *British Medical Journal* **303**: 1589–1591.
7. *The Complete Letters of Vincent van Gogh*. 2nd ed., 1978. Boston: New York Graphic Society.
8. Hemphill RE. 1961. The illness of Vincent van Gogh. *Proceedings of the Royal Society of Medicine* **54**: 1083–1088.
9. Tralbaut ME. *op. cit.* pp177–178.
10. Tralbaut ME. *op. cit.* p 276.
11. van Gogh-Bonger J. 1978. Memoir of Vincent van Gogh. vol I, pp XLV–XLVI, in: *The Complete Letters of Vincent van Gogh*. 2nd ed. Boston: New York Graphic Society.
12. Stokvis BJ. 1889. Over twee zeldzame kleurstoffen in urine van zieken. *Weekblad van het Nederlandsch Tijdschrift voor Geneeskunde* **2**: 409–417.
13. With TK. 1980. A short history of porphyrins and the porphyrias. *International Journal of Biochemistry* **11**: 189–200.
14. Waldenström J. 1957. The porphyrias as inborn errors of metabolism. *American Journal of Medicine* **22**: 758–773.
15. Tishler PV, Woodward B, O'Connor J, Holbrook DA, Seidman LJ, Hallett M & Knighton DJ. 1985. High prevalence of intermittent acute porphyria in a psychiatric patient population. *American Journal of Psychiatry* **142**: 1430–1436.
16. Bonkovsky HL, Sinclair PR, Emery S & Sinclair JF. 1980. Seizure management in acute hepatic porphyria: risks of valproate and clonazepam. *Neurology* **30**: 588–592.
17. Ivers WM. 1951. Acute intermittent porphyria. *Journal of Urology* **66**: 816–818.
18. Grossfeld E. 1951. Acute porphyria with unusual features. *British Medical Journal* Vol I for 1951: 1240–1241.
19. Sack GH. Jr. 1990. Acute intermittent porphyria. *Journal of the American Medical Association* **264**: 1290–1293.
20. Kappas A, Sassa S, Galbraith RA & Nordmann Y. 1989. The porphyrias. pp 1305–1365, in: *The Metabolic Basis of Inherited Disease*. 6th ed. CR Scriver, AL Beaudet, WS Sly & D Valle (eds.) New York: McGraw-Hill.
21. Laiwah AACY, Mactier R, McColl KEL, Moore MR & Goldberg A. 1983. Early-onset chronic renal failure as a complication of acute intermittent porphyria. *Quarterly Journal of Medicine* **52**: 92–98.
22. Lai CW, Hung TP & Lin WSJ. 1977. Blindness of cerebral origin in acute intermittent porphyria. *Archives of Neurology* **34**: 310–312.

23. Ridley A. 1969. The neuropathy of acute intermittent porphyria. *The Quarterly Journal of Medicine* **38**: 307–333.
24. Moore MR. 1980. International review of drugs in acute porphyria. *International Journal of Biochemistry* **12**: 1089–1097.
25. Welland FH, Hellman ES, Gaddis EM, Collins A, Hunter GW Jr. & Tschudy DP. 1964. Factors affecting the excretion of porphyrin precursors by patients with acute intermittent porphyria. I. The effect of diet. *Metabolism* **13**: 232–250.
26. Knudsen KB, Sparberg M & Lecocq F. 1967. Porphyria precipitated by fasting. *New England Journal of Medicine* **277**: 350–351.
27. Lip GYH, McColl KEL, Goldberg A & Moore MR. 1991. Smoking and recurrent attacks of acute intermittent porphyria. *British Medical Journal* **302**: 507.
28. Bickers DR, Miller L & Kappas A. 1975. Exacerbation of hereditary hepatic porphyria by surreptitious ingestion of an unusual provocative agent, a mouthwash preparation. *New England Journal of Medicine* **292**: 1115–1116.
29. Goldberg A & Rimington C. 1962. *Diseases of Porphyrin Metabolism.* pp 64–109. Springfield: C.C. Thomas.
30. Battersby AR, Fookes CJR, Matcham GWJ & McDonald E. 1980. Biosynthesis of the pigments of life: formation of the macrocycle. *Nature* **285**: 17–21.
31. Lascelles J. 1960. The synthesis of enzymes concerned in bacteriochlorophyll formation in growing cultures of *Rhodopseudomonas spheroides. Journal of General Microbiology* **23**: 487–498.
32. Scholnick PL, Hammaker LE & Marver HS. 1969. Soluble hepatic δ-aminolevulinic acid synthetase: end-product inhibition of the partially purified enzyme. *Proceedings of the National Academy of Sciences (USA)* **63**: 65–70.
33. Tschudy DP, Welland FH, Collins A & Hunter G Jr. 1964. The effect of carbohydrate feeding on the induction of δ-aminolevulinic acid synthetase. *Metabolism* **13**: 396–406.
34. Sassa S, Bradlow HL & Kappas A. 1979. Steroid induction of δ-aminolevulinic acid synthase and porphyrins in liver: structure-activity studies on the permissive effects of hormones on the induction process. *Journal of Biological Chemistry* **254**: 10011–10020.
35. Strand LJ, Felsher BF, Redeker AG & Marver HS. 1970. Enzymatic abnormality in heme biosynthesis in acute intermittent porphyria: decreased hepatic conversion of porphobilinogen to porphyrins and increased δ-aminolevulinic acid synthetase activity. *Proceedings of the National Academy of Sciences (USA)* **67**: 1315–1320.
36. Magnussen CR, Levine JB, Doherty JM, Cheesman JO & Tschudy DP. 1974. A red cell enzyme method for the diagnosis of acute intermittent porphyria. *Blood* **44**: 857–868.
37. Becker DM & Kramer S. 1977. The neurological manifestations of porphyria: a review. *Medicine* **56**: 411–423.
38. Müller WE & Snyder SH. 1977. δ-Aminolevulinic acid: influences on synaptic GABA receptor binding may explain CNS symptoms of porphyria. *Annals of Neurology* **2**: 340–342.
39. Meyer UA, Strand LJ, Doss M, Rees AC & Marver HS. 1972. Intermittent acute porphyria: demonstration of a genetic defect in porphobilinogen metabolism. *New England Journal of Medicine* **286**: 1277–1282.
40. Litman DA & Correia MA. 1985. Elevated brain tryptophan and enhanced 5-hydroxytryptamine turnover in acute hepatic heme deficiency: clinical implications. *Journal of Pharmacology and Experimental Therapeutics* **232**: 337–345.

41. Tschudy DP, Valsamis M & Magnussen CR. 1975. Acute intermittent porphyria: clinical and selected research aspects. *Annals of Internal Medicine* **83**: 851–864.
42. Meyer UA & Schmid R. 1978. The porphyrias. pp 1160–1220, in: *The Metabolic Basis of Inherited Disease.* 4th ed. JB Stanbury, JB Wyngaarden and DS Fredrickson (eds.) New York: McGraw-Hill Book Company.
43. Lieber CS, Teschke R, Hasumura Y & Decarli LM. 1975. Differences in hepatic and metabolic changes after acute and chronic alcohol consumption. *Federation Proceedings* **34**: 2060–2074.
44. Granick S. 1966. The induction *in vitro* of the synthesis of δ-aminolevulinic acid synthetase in chemical porphyria: a response to certain drugs, sex hormones, and foreign chemicals. *Journal of Biological Chemistry* **241**: 1359–1375.
45. Mustajoki P & Koskelo P. 1976. Hereditary hepatic porphyrias in Finland. *Acta Medica Scandinavica* **200**: 171–178.
46. Barker LF & Estes WL. 1912. Family hematoporphyrinuria in association with chronic gastroduodenal dilation, peculiar fits and acute polyneuritis: a preliminary report. *Journal of the American Medical Association* **59**: 718–719.
47. Waldenström J. 1937. Studien ueber Porphyrie. *Acta Medica Scandinavica* **82** (suppl.): 1–254.
48. Gates RR. 1946. *Human Genetics.* New York: Macmillan.
49. Wang A-L, Arredondo-Vega FX, Giampietro PF, Smith M, Anderson WF & Desnick RJ. 1981. Regional gene assignment for human porphobilinogen deaminase and esterase-A$_4$ to chromosome 11q23→11qter. *Proceedings of the National Academy of Sciences (USA)* **78**: 5734–5738.
50. Mustajoki P & Desnick RJ. 1985. Genetic heterogeneity in acute intermittent porphyria: characterisation and frequency of porphobilinogen deaminase mutations in Finland. *British Medical Journal* **291**: 505–509.
51. Kappas A, Bradlow HL, Bickers DR & Alvares AP. 1977. Induction of a deficiency of steroid Δ^4-5-α-reductase activity in liver by a porphyrin gene. *Journal of Clinical Investigation* **59**: 159–169.
52. Becker B, Shin DH, Palmberg PF & Waltman SR. 1976. HLA antigens and corticosteroid response. *Science* **194**: 1427–1437.
53. Herrick AL, McColl KEL, Wallace AM, Moore MR & Goldberg A. 1990. Elevation of hormone-binding globulins in acute intermittent porphyria. *Clinica Chimica Acta* **187**: 141–148.
54. Grandchamp B, de Verneuil H, Beaumont C, Chretien S, Walter O & Nordmann Y. 1987. Tissue-specific expression of porphobilinogen deaminase. Two isoenzymes from a single gene. *European Journal of Biochemistry* **162**: 105–110.
55. Grandchamp B, Picat C, Kauppinen R, Mignotte V, Peltonen L, Mustajoki P, Roméo PH, Goossens M & Nordmann Y. 1989. Molecular analysis of acute intermittent porphyria in a Finnish family with normal erythrocyte porphobilinogen deaminase. *European Journal of Clinical Investigation* **19**: 415–418.
56. Grandchamp B, Picat C, Mignotte V, Wilson JHP, Te Velde K, Sandkuyl L, Roméo PH, Goossens M & Nordmann Y. 1989. Tissue-specific splicing mutation in acute intermittent porphyria. *Proceedings of the National Academy of Sciences* (USA) **86**: 661–664.
57. Goldberg A. 1985. Molecular genetics of acute intermittent porphyria. *British Medical Journal* **291**: 499–500.
58. Tralbaut ME. *loc. cit.* pp. 16–17.

Chapter 6
Other Hypotheses

Figure 6.1. Vincent van Gogh: *Self-Portrait*, 1887, Paris, oil on canvas, 46 × 38 cm (18.1 × 15″), Österreichische Galerie, Vienna

Diseases exist to remind us that we are not made of wood, and it seems to me this is the bright side of it all.
Vincent to Monsieur and Madame Ginoux, letter 622a, from St. Rémy, January, 1890.

A typical newspaper article on Vincent starts by telling the audience that there are one hundred and one diagnoses on van Gogh's illness. This is followed, explicitly or implicitly, by the question, "when will they get it right?" Actually, there are no more than a dozen serious proposals, but many renditions. I started with the assumption that all the authors were sincere but found that only a few advanced the field. Unfortunately, the majority of hypotheses were loosely conceived and poorly documented, but they landed in the literature and to this day are still quoted without much evaluation.

The reasons for discussing other hypotheses are threefold. First, there is a need to establish a differential diagnosis between acute intermittent porphyria[1] and other diseases and disorders. Second, overlap will become apparent within certain areas, and it is of more than passing interest to explore how much of the total symptomatology can be accommodated by other diagnoses. And third, even though some speculations do nothing more than expand the mythology about Vincent, they still need be dismissed to the realm of idle conjecture by offering some solid scientific criticism.

My approach is to introduce a particular working hypothesis, place the unique terminology in historic perspective, provide a modern description of the disease or disorder under discussion, examine the salient points in the light of Vincent's letters, and evaluate the worth and contribution of that particular hypothesis to our overall understanding of the subject. The more extensive compilations of hypotheses on van Gogh's illness are found in Beer,[2] Perry,[3] Tralbaut,[4] and Lubin.[5] A good place to start is the 1889 register (number 4, page 142) of the *Maison de Santé de Saint-Rémy-de-Provence* which is reproduced in Tralbaut.[4]

The asylum register

Columns 1 and 2 contain the patient's name, birthplace, and the notation that he was *committed by his brother.* (Italics here and below are mine.) This flies in the face of the popular view that Vincent was self-committed. The commitment was *voluntary* but was made by Theo who was required to seek release of Vincent the next year. Column 3 contains a transcription by Dr. Peyron of a note from the Hospital at Arles (written by Dr. Urpar and dated the previous day, i.e. May 7) to the effect that Vincent was "stricken six months ago by *acute mania with generalized delirium*. ... although his condition is greatly improved he feels that it is to his benefit to be treated in a mental asylum." Column 4, written 24 hours after admission, i.e. on May 9, states that when the patient was in Arles he was "stricken by *acute mania with hallucinations of his sight and hearing,* which led him to mutilate himself. ... [at present] he appears to have regained his reason, but he feels neither the strength nor courage to live at liberty ... I [Dr. Peyron] believe ... that Monsieur van Gogh is subject to *epileptic fits, at very great intervals,* and that there is reason to put him under extended observation at this establishment [St. Rémy]." Below this, in the same column, appears a report, dated May 25, which essentially restates the above and goes on to say that the patient "has experienced a perceptible improvement in his condition, but there is reason to keep him in this establishment to continue his treatment."

In column 5, on another page, is written the admission date of May 8, 1889. Column 6, designated "monthly notes by the asylum physician" is again repetitious but adds that the hallucinations *"terrified"* the patient and "he retains only a *vague memory* of [the ear cutting affair] and *cannot explain* it. He tells us that one of his mother's sisters was *epileptic,* and that there are several cases in the family. What happened to this patient would only be the continuation of what has happened to several members of his family. ..."

Column 7, headed "Orders from the prefect or mayor," and column 8, reserved for a "Transcription of an interdiction certificate and the name of the guardian," are vacant. In column 9 the date of exit is given as May 16, 1890. Column 10, "Observations," contains only one word *Guérison* (cured) and is signed Dr. T. Peyron. That's it! Never has so little been written about so much.

Dr. Urpar used the term "acute mania" which is now taken to mean a mood disorder characterized by expansiveness, elation, agitation, hyperexcitability. He also used "delirium" which is defined as an acute, reversible, organic mental disorder characterized by reduced ability to maintain attention to external stimuli, disorganized thinking manifested by rambling or incoherent speech, a disorientation to time place or person, and memory impairment. Delirium was used to characterize Vincent's condition in letters by Dr. Rey and by Vincent himself, and it certainly conforms with the written descriptions by his friend Roulin and the Reverend Salles. By "generalized" delirium I suppose that Dr. Urpar meant "complete" although this combination is not in current usage.

In the entry that followed, Dr. Peyron substituted "hallucinations" for "delirium." It would have made more sense if he had added rather than substituted because the meaning is now quite different; an hallucination is a sense perception without a source in the external world, i.e. hearing, seeing, or smelling something that isn't there. It has been assumed that auditory and visual hallucinations were items that Vincent conveyed to Dr. Peyron during his first interview. That Dr. Urpar did not mention them has been interpreted by Tralbaut[4] as evidence of meager involvement by the Director at Arles compared with the attending, Dr. Rey. The key words in the St. Rémy register are *mania, delirium,* and *hallucinations;* the leap to "epilepsy" is my point of departure with Dr. Peyron. Nonetheless, the diagnosis of epilepsy has the longest tenure and must be considered.

Epilepsy

Epilepsy is defined as a paroxysmal (sudden and recurring) transient disturbance in brain function that is manifested by episodic impairment or loss of consciousness, abnormal motor phenomena, psychic or sensory disturbances, or perturbation of the autonomic nervous system. The derivation of the word is Greek; it means seizure. In medicine, the first meaning of seizure is a sudden attack or recurrence of a disease. The second, and our present context, implies an attack of epilepsy. Accordingly, the term *epileptic seizures* is redundant, but common parlance. Another basic term is *convulsion,* which means a violent involuntary contraction, or series of contractions, of the normally voluntary muscles.

Several diseases and conditions are complicated by seizures. Convulsions appear after withdrawal from alcohol or barbiturates. Uremia, abnormal amounts of urea in the blood, may be tolerated for a few days and then cause a rapid onset of twitching, trembling, myoclonic (shocklike) jerks, and generalized seizures. Other acute illnesses which present with seizures include brain tumors, hyponatremia (low level of sodium in the blood), thyrotoxicosis (excessive quantities of thyroid hormones), porphyria, and hypoglycemia (low blood glucose). Lead and arsenic are the most frequently encountered metallic intoxications which cause convulsions. Niedermeyer[6] emphasized that *epilepsy is not a disease* but rather an abnormal reaction of the brain due to numerous causes. The following is a classification of epileptic seizures based on the 1981 international proposal.[7,8]

I. Partial seizures (focal, local)
    A. Simple (consciousness not impaired)
    B. Complex partial seizures (impairment of consciousness)
    C. Partial seizures which evolve into generalized seizures

II. Generalized seizures (convulsive or non-convulsive)
    A. Absence seizures

B. Myoclonic seizures (a subtle or massive muscle contraction)

C. Clonic seizures (more prolonged and rapidly successive myoclonic events)

D. Tonic seizures (characterized by rigid stretching of the body and extension and abduction of the extremities)

E. Tonic-clonic seizures (tonic followed by clonic convulsions)

F. Other combinations

III. Unclassified epileptic seizures. (This includes several neonatal types which cannot be classified because of incomplete data)

An older term still in clinical and popular usage is *grand mal,* which includes but is not restricted to tonic-clonic seizures. This is one of the classical, maximal epileptic responses of the brain. A patient previously lying at rest will suddenly raise and extend both arms, and hold them stiffly except for clenching and unclenching the fists; this is the tonic stage. Then the extremities, as well as the facial muscles, start constantly jerking and twitching; this is the clonic phase. Grand mal is probably the most common of all seizure types. It will be recalled from chapter 4 that absinthe abuse induced convulsions that mimicked epilepsy except that the order of presentation was usually clonic followed by tonic. Also, it should be mentioned that generalized convulsions appear prominently during a withdrawal period in patients addicted to alcohol or barbiturates.

Petit mal seizures (absence seizures) refer to a brief lapse in consciousness, usually no longer than 20 seconds. The facial expression during the ictal (the actual seizure) period is blank but rhythmical eye movements may attend; the postictal (after the seizure) events are minimal and there is usually a prompt return to the normal condition. Accordingly, if a patient has fatigue after a seizure (or evidence of any abnormal postictal state), this virtually rules out petit mal seizures.

Complex partial seizures (or partial seizures with complex symptomatology) present with a great variety of features, but they invariably involve a marked loss of consciousness and are the major type of uncontrolled seizure seen in adults. This sub-category encompasses several older terms such as *psychomotor* seizure and *temporal lobe* seizure.[9] Complex partial seizures are often preceded by an *aura,* a subjective sensation or motor phenomenon that marks the onset of an attack. The aura is usually brief, is itself a simple partial seizure, and may take various forms including a "breezy" feeling (as the Latin and Greek derivations imply) or a false sensation of movement. As the seizure develops the patient often exhibits certain automatic but bizarre behaviors such as chewing and swallowing movements, fumbling with clothing, posturing, thrashing of arms and legs, and so on.

There is no single cause for complex partial seizures. The only common denominator is the anatomical structure which is affected. It is the deep regions of the temporal lobe, and connecting limbic structures, which undergo electric discharge. Confirmation comes from electrical stimulation via depth electrodes in the temporal lobe, whereby many of the phenomena noted during spontaneous seizures are reproduced.

Evidence of a genetic factor in epilepsy has accrued from the study of those inherited diseases in which epilepsy is a symptom, from the breeding experience with certain strains of experimental animals, and indirectly from familial patterns.[10] A significant fraction of seizure patients, otherwise unselected, are found to have a positive family history for this abnormal reaction.

Vincent and epilepsy?

There is much evidence from the letters to indicate that Vincent's medical crises were extremely debilitating. Time and again he wrote to Theo about fatigue and malaise after an attack [see chapter 3]. Petit mal seizures are certainly not indicated. Likewise, grand mal seizures have never received much diagnostic support because tonic-clonic convulsions were not described by Vincent or his doctors. Thus the classical sorts of epilepsy, that were well understood in Vincent's time, were hardly indicated. For this reason I agree with Tralbaut[4] that Dr. Peyron's unqualified diagnosis of "epilepsy" was based upon the patient's preconceived view.

When Vincent said, "I am a madman or an epileptic" (letter 589) he undoubtedly preferred the latter, although I would say he was neither. Nonetheless, many psychoanalysts felt comfortable with epilepsy and embraced this diagnosis while they looked for hidden messages in Vincent's paintings.[11,12] Jaspers,[13] in 1922, opined schizophrenia while gazing at landscapes, without even reading the St. Rémy register. Riese[14] in 1925, and Minkowska[15] in 1933, rejected schizophrenia but embraced epilepsy as if nothing else were remotely possible.

As early as the 1870's, Hughlings Jackson had described certain hallucinations with seizures that he related to a pathologic condition of the temporal lobe.[16] It was not until the 1930's that so-called "psychomotor" seizures were well described.[17] In the 1950's the anatomical adjective "temporal lobe" was again preferred, even though some other parts of the brain were sometimes involved.[18] Today, these are all lumped under complex partial seizures.[19]

Dr. Edgar Leroy who worked at St. Rémy Asylum, albeit many years after van Gogh's sojourn, and Dr. Victor Doiteau considered that Vincent was epileptic but found no evidence of aura or frank convulsions.[20] They suggested temporal lobe epilepsy, and several authors[4, 21, 22] subsequently opted for this diagnosis. Pickvance[23] was typical of those who wished to be less specific and called Vincent's illness "some sort of epilepsy."

A diagnosis of temporal lobe epilepsy might explain Vincent's hallucinations, the episodic nature of his illness, and the interictal periods of normalcy. However, the usual time course of minutes or hours that attends the various forms of complex partial seizures does not fit the days and weeks of Vincent's crises. A rare condition called *complex partial status* would have to be invoked; it usually presents as stereotypic purposeless movements with gross impairment in coping with daily activities, and does not fit Vincent's case. More importantly, epilepsy does not accommodate the numerous gastrointestinal complaints. Likewise, some of the factors which exacerbated his illness such as malnutrition and fasting are not noted for inducing temporal lobe epilepsy. The age of onset datum provides no particular support for complex partial seizures as it does for acute intermittent porphyria. While generalized seizures are sometimes precipitated by alcohol abuse, this is not clear for complex partial seizures, which neither gains nor loses merit by the fact that some of Vincent's crises followed trips to Arles.

Drug therapy in the 1880's was limited, but Vincent's fits and confusion (letter W11) seem to have been controlled in Arles by bromide (letter 574), which would be indicated for absinthe intoxication or porphyria, but not for temporal lobe epilepsy. Bromides are effective against grand mal and simple partial seizures but not for complex partial seizures.[6, 24] Monroe[25] noted that the limbic system is exquisitely sensitive to stress and external toxins including alcohol and remarked on Vincent's affinity for absinthe. He joined Hemphill[24] in suggesting that a toxic psychosis was an important part of the illness picture. How much of the symptomatology can be explained by alcohol alone?

Alcoholic seizures and delirium tremens

A sustained period of imbibing alcohol is the underlying factor, but the following symptoms and course of illness are only manifested during a period of relative abstinence or after cessation.[26] The most common initial complaint is "the shakes," a state of extreme tremulousness, which may be combined with nausea and vomiting. This usually happens in the early morning following days of drinking (after a short abstention during sleep) and is calmed by more alcohol. This pattern may be repeated for days or weeks and then the subject stops drinking for financial or other reasons.

The symptoms which start in the next day or so, after complete cessation of alcohol intake, are even more dramatic. The patient is alert but startles easily; he may exhibit tachycardia, nausea and retching; he suffers from insomnia and is preoccupied with his miserable state. He becomes mildly disorientated with respect to time and has poor memory. The tremor may be so violent that the patient cannot stand, feed himself, or speak clearly. It may take two weeks before he can achieve full composure and sleep without sedation.

About one-quarter of these tremulous patients have nightmares. They also suffer hallucinations both auditory and visual. Imagined voices seem to be coming from behind doors and are often attributed to absent family members. Interestingly, these imagined sounds elicit appropriate responses by the patient who is not confused or disoriented as before, but now has an intact memory. Hallucinations may recur intermittently for days until an improvement is signalled by the patient doubting their reality. At this stage the patient may exhibit symptoms resembling schizophrenia such as illogical thinking, vagueness, and a disassociation of externally expressed emotion from thought content.

In this same setting, within one-half to two days after drinking has ceased, seizures are also common. These seizures are usually of the grand mal type. Victor and Adams[26] call them "rum fits" and state that the electroencephalographic data reflect a sequence of electrical discharges definitely induced by alcohol abuse rather than "latent epilepsy" manifest by alcohol. Almost a third of patients with grand mal seizures go on to develop delirium tremens. Several settings are possible, but to continue our generic case the patient may be recovering from several days of tremulousness, hallucinosis, and one or more seizures, when he suddenly develops delirium tremens. This is characterized by profound confusion, delusions, vivid hallucinations, tremor, agitation and sleeplessness, dilated pupils, tachycardia, and profuse perspiration.[26] Delirium tremens ends as abruptly as it begins in the majority of cases. The patient falls into a deep sleep and awakens lucid. There may be one or more lapses. It can last days or as long as two months. About 15% of cases end fatally.

In chapter 4, I abstracted a case of absinthism which certainly included many of the symptoms just described for alcohol withdrawal syndrome. Absinthe abuse, according to Magnan and other investigators, was notable for the induction of hallucinations *prior to or even without tremulousness*. Also, active heavy drinkers of absinthe could experience these symptoms without going through the characteristic withdrawal period associated with alcohol alone. It has been tacitly assumed by most commentators that Vincent's doctors in Arles and St. Rémy would know alcoholism when they saw it! In chapter 3, I quoted and referenced letters by Vincent about alcohol – it is worth repeating that Dr. Rey in Arles certainly advised Vincent to cut back on his drinking (see for example letters 581and 581a).

We do not know the extent of Vincent's drinking, but we do know that he admitted to excesses. It is again assumed that the hospital in Arles and the asylum at St. Rémy endeavored to restrict alcohol consumption; how successful they were is open to question. I am convinced that Vincent engaged in "social" drinking when he visited friends in Arles, but this was for a relatively short time of a day or so. The time course of his illness, and the duration of some of the crises in the asylum, do not fit alcohol withdrawal syndrome per se. I believe it was more of a sensitivity to alcoholic beverages than an extraordinary dose. Alcohol is a an exacerbating factor for acute intermittent porphyria. It can also be a factor in the disease called manic-depressive psychosis.

Manic-depressive (bipolar) illness

It is worth asking what Dr. Peyron meant when he entered *acute mania* in the St. Rémy register in 1889. In the classical world, *mania* was a general expression for madness, but had overtones of rage, aggression and excitement.[27] By 1800, it still meant "madness" but was also the best example of "total insanity." In 1900, it assumed its present psychiatric meaning of a mood disorder characterized by expansiveness, elation, agitation, hyperexcitability, hyperactivity, and increased speed of thought and speech (flight of ideas). Thus the bulk of this interesting evolution took place in the nineteenth century.

In Pinel's book (1818), mania was a disorder of one or more faculties with sad, gay, extravagant or raging affect, but always included blind aggression.[28] This concept was further modified, and then fell out of usage for a time as it was replaced by *monomania* (preoccupation with one subject or idea). This substitution had a limited tenure, but when mania returned to vogue the term now had a new meaning. It referred to a particular category.

Meanwhile, *melancholia* was also undergoing an evolution. Up until the beginning of the nineteenth century, the prime meaning was intensity of idea, the image of the mind being strongly fixed on, and frequently returning to, a single set of ideas, to an extent that was deemed unhealthy. The connotation of sadness was not always present, and many forms of behavior that have little relationship (from our perspective) were included in the general class of melancholia. Not surprisingly there was even a "productive melancholia" that today might be more akin to intense, creative, concentrated thinking directed at a particular problem, while excluding all day-to-day distractions. (This is supposed to be the hallmark of creative individuals!) In 1820, Esquirol suggested that melancholia would be better left to the language of poets and replaced it with stricter medical language.[29]

In the 1850's the older notions of mania and melancholia were replaced by mania (with the new meaning), and depression. This change was ushered in by Falret, in 1854, when he described so-called "circular" insanity in which mania and melancholia alternated at regular intervals.[30] Note that the term melancholia was still used, but the meaning was now approaching depression. The same year, Baillarger[31] also wrote about these two states, but also included an intercalated period of normalcy as part of the syndrome. A protracted dispute over priority ensued, although it would seem that Baillarger's "double-form" disease was closer to our present concept[32] of manic-depressive psychosis or bipolar disorder. It should be mentioned in passing that Dr. Paul Gachet attended lectures by both Falret and Baillarger.

The title of Dr. Gachet's thesis was *Étude sur la Mélancolie.* The work was written in 1858, in the middle of this transition period in terminology. His thesis was really a compendium of principles for moral treatment of the insane, spiced with a philosophical vitalism that he encountered at the Montpellier Medical School.[33] Gachet felt that melancholia was a pathological sadness of such a degree that it inhibited the patient's ability to carry out the normal activities of everyday life. His thesis title was already somewhat old-fashioned, nonetheless his concept of melancholia was closer to our present *depression.*

The assumption made by some commentators that manic-depressive psychosis was unknown in Vincent's day is incorrect. The French Academy of Medicine had major meetings on the subject starting in 1880. How well it was recognized, received, or dealt with in Arles and St. Rémy in 1889–90 is an open question, so we are still left in some doubt as to the intent of Dr. Peyron when he talked about mania. I am inclined to think that he was referring to the events which surrounded the ear-cutting incident, and the complaints of neighbors about Vincent's drinking sprees, which led to his readmission to the Arles hospital in 1889. If that is true it was "old fashioned" mania à la Pinel.

Mania (the modern concept)

The major signs and symptoms[34] demonstrate a disorder of affect, i.e. an improper external expression of emotion attached to ideas, which is manifested as exaggerated irritability, hyperactivity, and an outgoing behavior of psychotic proportions. The mood is one of excitement, elation, being driven with a special mission, leading the pack in laughter and emotional outbursts, assuming control of events. Such patients usually sleep less. They awake alert and ready to go. However, if the manic phase is prolonged there follows inevitably an appearance of fatigue and the emergence of exhaustion, all in a setting of increased irritability. Under these conditions, underlying aggression may suddenly be released.

Manic excitement produces rapid and disorganized speech (flight of ideas). Sometimes the patient has grandiose ideas, delusions of grandeur concerning position, power, wealth or connections. Drinking of alcoholic beverages, nonmedical use of pharmacological agents, and neglect of regular eating habits are frequently observed in manic patients. In combination with hyperactivity, this leads to weight loss.

About one-third of manic patients are fully aware of being ill. They can monitor their effect on other people and can pull themselves together in order to talk logically and convincingly for at least a short time. Very active, delusional, and agitated subjects lack this intellectual insight. Patients who present with only manic attacks do better prognostically than patients with both depressive and manic episodes. The average duration of an acute manic episode is approximately two to four months but can be as long as nine months.

Depression, the modern concept

Depressed manic-depressive patients suffer from a sense of being overwhelmed from inside by symptoms they are helpless to control.[34] Disturbed sleep and insomnia, low self-esteem, and an obsessional preoccupation with the body and its functioning are all prevalent. They feel shattered, devalued, unhappy; their facial expressions and manners reflect this. They are convinced their state is hopeless and will never change. Sawer-Foner[35] calls it "the delusion of the eternity of the depressed state," and claims that all psychotically depressed patients, bipolar or unipolar, demonstrate this delusion.

From a diagnostic viewpoint, manic episodes constitute important criteria for manic-depressive illnesses. On the average there are nine to ten depressive episodes for every manic episode. A course of regular cycling between mania and depression, which is popularly held, is rarely observed. First attacks of mania usually occur before age 30; first attacks of depression are more prevalent after 35. A histogram of overall frequency versus age-of-onset for manic-depressive patients [n = 898] peaked with the 15–19 year group, and was closely followed by the 20–24 year group.[36]

Delusions and hallucinations

Delusions are defined as false beliefs that are firmly maintained in spite of evidence to the contrary and the fact that other members of the culture do not share the belief. An hallucination is a sense perception without a source in the external world; auditory and visual hallucinations are the most common. For manic depressives the themes of their delusions and hallucinations usually reflect the dominant mood of the acute episode, i.e. either manic or depressive. Mood-incongruent psychotic symptoms, especially delusions of grandeur, were once thought to be more prevalent in schizophrenia but there is no firm base for this assumption.[37]

The rate of suicidal ideation is high among depressed patients. The study of Roose et al.[38] found a significantly higher rate of suicide associated with hospitalized, depressed patients who had delusions. About 9% of manic-depressive patients commit suicide, mostly in the first decade of their illness. Some are "unintentional suicides" because they tend to be risk-takers. Alcohol and drugs may play a role in risks that go wrong.

The response to new drugs has often compounded difficulties in establishing meaningful diagnostic tests for manic-depressive psychosis. Often it has led to a seemingly unfortunate splitting of cases into subcategories. Notwithstanding intense research investigations in the last few decades, the biochemical marker for manic-depressive psychosis has yet to be discovered. I believe, along with many others, that a chemical description for this debilitating psychosis will eventually be forthcoming. At that time the symptomatology of this great "catch-all" of a diagnosis will eventually yield to more concrete explanations.

Vincent van Gogh and manic-depressive psychosis?

The letters of Vincent van Gogh contain numerous accounts of episodes in which he displayed bursts of activity, when he painted furiously, did not eat properly, took little care of himself, and found relaxation only with "a lot of drinking or heavy smoking" (letter 507). In the early months at Arles he mentions, "Some days I still suffer from unaccountable, involuntary fits of excitement or else utter sluggishness, but that will pass as I get calmer" (letter 492). And then there were periods of utter exhaustion: "I have been and still am nearly half dead from the past week's work ... I have just slept sixteen hours at a stretch, and it has restored me considerably" (letter 553). There are references to melancholy (see chapter 3) or to moods that could be taken as depression, "My brain is still feeling tired and dried up, but this week I am feeling better than during the previous fortnight" (letter 558b).

The reassuring words of Dr. Rey to Theo "the overexcitement has been only temporary" (letter 567) following Vincent's first crisis could be interpreted as a reference to the older definition of mania. "I hope I have just had simply an artist's fit" was Vincent's optimistic view (letter 569). But in the next letter he was suffering from insomnia.

In letter 573 Vincent gives us a detailed picture: "During my illness I saw again every room in the house at Zundert [his birthplace], every path, every plant in the garden ... down to a magpie's nest in a tall acacia in the graveyard." This seems benign but in the next letter he spoke of the "unbearable hallucinations." Unfortunately, these were never described in the published letters except to say that "the attacks [in St. Rémy at least] tend to take on an absurd religious turn" (letter 605). In February 1889, Vincent was taken to the hospital at Arles, he imagined that people wanted to poison him (footnote to letter 576) and two weeks later he admitted to feeling "completely out of sorts" (letter 577). In mid-March he feared that he might "easily relapse into a state of overexcitement" (letter 579). By the beginning of April he was well "except for a certain undercurrent of vague sadness" (letter 583). He had thoughts of suicide: "If I were without your friendship, they would remorselessly drive me to suicide, and however cowardly I am I should end by doing it" (letter 588).

In the St. Rémy asylum initially he was "not so unhappy ... [but] I have not yet reached the point where I ought to think of leaving here; I should have this depression anywhere" (letter 592). The chronology of successive attacks need not be repeated here (see chapters 2 and 3). There were "abominable nightmares" and "great depression sometimes" (letter 613). By the end of January 1890, he told his friend John Russell about his "serious nervous crises and delirium" (letter 623a). Then he desperately needed a change and begged to come north again because "the whole horrible attack has disappeared like a thunderstorm" (letter 633). In Auvers he showed the first signs of depression in letter 636: "I can do nothing about my disease. I am suffering a little just now." He also reported that Dr. Gachet felt that "if the melancholy or anything else became too much for me to bear, he could easily do something to lessen its intensity" (letter 637). The march to suicide has been documented (chapter 2) and see also chapter 10.

These abstractions can be used to support manic-depressive psychosis. However, there were recurring themes about malnutrition and alcohol abuse which kept cropping up in Vincent's letters; the concept that if he could afford to, or elected to, eat better then his problems might disappear. When he wrote that "a more violent attack may forever destroy my power to paint" (letter 605) he signalled the possibility of a massive irreversible insult for which his body was ill prepared. Dr. Rey urged him on several occasions to eat more and better food, and to cut back on drinking and smoking. The age-of-onset of Vincent's crises in the last two years is somewhat later than average for manic-depressive psychosis but a large range is possible, and there is always a fuzzy uncertainty of first problems with this disease.

Perry (1947) was probably the first to seriously consider manic-depressive psychosis as a diagnosis for Vincent van Gogh; her expression was "cyclothymic personality with episodes of depression and mania."[3] Others have followed, and today it is one of the most commonly touted working hypotheses. Bipolar affective disorder is compatible with creativity as evidenced by the long lists given by Andreasen and Glick[39] and Goodwin and Jamison,[36] of many examples of undoubtedly creative people together with some indication of their suffering from that disease. (There is also some overlap in those series with Dr. Gachet's thesis list of outstanding individuals who suffered from "melancholia.") But this circumstance is hardly unique to manic depressive psychosis; for example, Monroe[25] has presented a forceful case for many of these same individuals having a propensity for limbic seizures and what he calls "brainstorms." Likewise, Goodwin[40] has shown that alcoholism is compatible with outstanding ability in writers.

In the opinion of Hemphill,[24] van Gogh was "a manic-depressive who developed confusional episodes and fits in the last two years of his life due to the toxic action of thujone, the active agent of absinthe." Hemphill's contribution in 1961 was twofold; he was the first to correctly refer to Vincent's "epilepsy" as a disorder rather than a disease, and he stressed the evidence for a toxic psychosis. He supposed that the gastrointestinal complaints came from the absinthe abuse. Vincent himself seemed to be approaching this idea when he wrote, "I have tried to compare the second attack with the first ... it seemed to be caused more by some outside influence than by something within myself" (letter 605). Loftus and Arnold[1] are convinced that it was the underlying illness of acute intermittent porphyria that made Vincent so sensitive to absinthe and malnutrition and was also the cause of his "stomach" problems.

Schizophrenia

Currently, schizophrenia is believed to comprise a large and common group of mental disorders, whose classification is still evolving.[41] They are a subcategory of functional psychoses and are recognized entirely by psychologic analysis rather than physical findings or laboratory tests. Important features include misinterpretation of reality, and sometimes frank delusions and hallucinations. Mood is usually flat, apathetic, or inappropriate to the situation. Behavior and appearance are often odd or even bizarre, and most authorities include a paranoid form.

The schizophrenic patient typically has an overwhelming lack of drive, enthusiasm, and assertiveness. There can be a sense of the body being detached from the mind, the idea that the surroundings are different and strange. The patient tends to confuse parts for wholes, lumping and condensing items in an illogical manner. Schizophrenics are often literal and lose the ability to think abstractly; they also have difficulty in separating the relevant from the irrelevant.

Hallucinations (especially auditory) and delusions occur in most patients with schizophrenia. Perceptual distortions also occur and are manifested in speech, writing, art – all with some common characteristics of excessively concrete symbolism, and elaborate systems of logic and language. Schizophrenia also leads to a reduction in the ability to perceive and remember faces and to observe expressions.[42]

Although Western painting over the last few centuries has prepared us for intentional deformation of faces in paintings and sculptures, in order to express emotion, the extraordinary preoccupation with this in some artists has sometimes been tied to schizophrenia. James Ensor's *The Entry of Christ into Brussels* has a great number of deformed faces, expressions of fear and anxiety, and figures depicted with masks; this has been interpreted, in 1988, as suggesting schizophrenia.[43] Alterations in face perception by documented schizophrenics are fully discussed and exemplified in the Prinzhorn collection.[44]

Progressive changes in content and style have thus been observed in the work of artists who are deemed to have schizophrenia.[44] The reverse – namely to see the psychosis in unknown artists by looking at their work – is obviously more difficult, but not sufficiently daunting to inhibit the proponents of schizophrenia for Vincent van Gogh. Such was the approach of Jaspers,[13] in 1922, who is still quoted under this heading. But the hypothesis has little credibility as can be immediately judged from reading the first two paragraphs of this section describing schizophrenia and comparing the features with details in the artist's vita (see chapter 2). Vincent had hallucinations, and he also had at least one episode of paranoia when he thought that neighbors were trying to poison him in Arles, but these are not specific for schizophrenia. Perry[3] remarked that "Vincent never withdrew from the world; he was cast out because of his behavior." Also, the progressive deterioration of the untreated schizophrenic is lacking in van Gogh. The schizophrenic has a decrease in affect whereas Vincent's letters and pictures were surcharged with emotion. Hemphill[24] saw no sign of schizophrenia in the artist and emphasized that there was never any phantasy formation, and that his letters were lucid and logical.

Sunstroke

Roch Grey[45] declared in no uncertain terms, "It is likely, sure indeed, that van Gogh suffered from a *chronic form of sunstroke* which manifested itself in his spasmodic attacks when he seemed to lose some of his reason. His violences which had the aspect of regular madness were only struggles between the conscious creative force of the man of genius and the gigantic unconscious force of the sun, which as brute matter and deprived of sensibility, could the better wage its attacks with purely mechanical means." This language surely epitomizes the romantic style of medical diagnosis. But we should not be too unkind to Grey who wrote this in 1924, in the introduction of his little book with 33 reproductions of Vincent's paintings. It was one of the first unbridled appreciations of van Gogh the artist, and we can excuse the author for being carried away.

However, Beer[2] took the idea seriously enough to give it passing mention in his thesis (1935). Later, in a popular article,[46] this author stated that "Dr. Gachet of Auvers-sur-Oise considered [a diagnosis of] turpentine poisoning and the effects of *too intense sun on a Nordic brain*." Again, no supporting evidence was provided. I have not been able to confirm the attribution to Dr. Gachet, but I assume some verbal anecdote that slipped into the van Gogh literature. Vincent himself remarked upon being "dazed with the sun" (letter 512) that "beats down on one's head ... [and] makes one crazy" (letter B15). Sunstroke results from uncontrolled elevation of body temperature after the reflex sweating system is overtaxed by environmental heat.[47] There is some indication that hallucinations can be induced by sunstroke,[48] and see also chapter 8. Vincent may have been a bit reckless in his exposure to the sun, but there was certainly more to his illness than heatstroke; the time course and the rest of the symptomatology can not be accommodated under this heading.

Syphilis

Syphilis can be acquired either congenitally or, most often, by sexual contact with an infected individual. The primary stage is remarkably free of systemic signs, the patient is entirely well and usually free of fever but, at about 1–12 weeks after contact, 50% of females and 70% of males develop a primary lesion (chancre) at the site of infection by the spirochete *Treponema pallidum*. In the secondary stage, at 2–12 weeks after the primary stage, a skin rash appears. Constitutional symptoms that may accompany secondary syphilis include fever, weight loss, malaise, and anorexia. There follows an asymptomatic latent stage that may last decades. About 30% of untreated patients go on to develop tertiary lesions but clinical disease occurs in only half of these cases; this fraction is 15% overall. About 80% of the tertiary lesions affect the cardiovascular system, 10% are chronic focal inflammations (gummas) in the liver and other sites, and up to 10% involve the central nervous system (neurosyphilis), i.e. 1.5% overall.[49]

Neurosyphilis

The major clinical categories of symptomatic neurosyphilis are meningovascular and parenchymatous syphilis. The latter includes tabes dorsalis, characterized by degeneration of the posterior columns of the spinal cord and posterior spinal roots. The interval from infection to expression of symptoms is about 27 years. The clinical presentation is characterized by lancinating lightning pains of short duration which occur most frequently in the extremities. Other symptoms include morbid or perverted sensation (paresthesia) and bladder disturbances, and a failure of muscular coordination (ataxia). Urinary retention and the complications of urinary tract infections are common. Ambulation is affected and a characteristic broad-based, stamping gait is displayed.

Another form of parenchymatous syphilis, general paresis of the insane, is associated with direct invasion of *T. pallidum* into the brain. For unknown reasons the syndrome is more common in males. The average interval from infection to onset of general paresis is 20 years. There is chronic inflammation, thickening and fibrosis of the meninges, extensive neuronal loss, and demyelination of white matter. The clinical presentation combines symptoms of neurologic and psychiatric illness. Early manifestations are irritability, memory loss, impaired cognition, and personality changes. As the disease progresses, confusion and delusional behavior are more apparent. Some patients have hallucinations or grandiose delusions. The patient becomes bedridden and is plagued by seizures. The course of the untreated disease is inexorably progressive, with death occurring sometimes in months.[50]

Vincent and neurosyphilis?

Neither the gamut of his symptoms nor the time course of his crises fits neurosyphilis. Vincent was treated for gonorrhea in The Hague in mid-1882 at age 29. He may have had a recurrence in Antwerp in 1885–86, at age 32, although Tralbaut's contact (see chapter 3) claims that his grandfather treated Vincent for syphilis. Even if he had contracted syphilis in The Hague, the major crises in Arles (age 35) would have been extraordinarily early for the onset of neurosyphilis, and his lengthy remissions from illness also negate the possibility. Mercury treatments were used at Arles and St. Rémy for syphilis, but Doiteau and Leroy[20] found no indication that Vincent received mercury.

Theo and neurosyphilis?

I reviewed what has been published about Theo van Gogh's health record in chapter 3. He died in Willem Arntsz Stichting, the mental institution in Den Dolder (Holland), on January 25, 1891. Some cryptic medical records from that institution were released in 1990 by a local historian to Dutch newspapers. The story, which covered the 38 days from Theo's move out of Paris to Den Dolder until his death, ends dramatically, "the final diagnosis was dementia paralytica (general paresis)."

Dementia paralytica was described by Bayle, as early as 1822. Quincke is credited with introducing the lumbar puncture procedure together with examination of the cerebral spinal fluid for spirochytes, in 1892. Today, a diagnosis would be based on serology of the cerebral spinal fluid, but this technology was not available until well into the twentieth century. General paresis was overly diagnosed in the nineteenth century and the psychiatric and neurological symptoms recorded for Theo's case are far from definitive. An autopsy examination could have provided confirming evidence but apparently was not performed. In any event, the time course of Theo's illness makes the case for neurosyphilis highly unlikely. Here is a recapitulation of the highlights given in chapter 3.

Theo took his first job away from home at age 15. He suffered a life-threatening illness at 19. At 25, he had bouts of illness with headaches and malaise. In 1886, his future brother-in-law, Andries Bonger, reported that Theo (then 29) was frightfully ill and, six months later, that Theo had been paralyzed. At 31, Theo was consulting Drs. Gruby and Rivet in Paris, and six months later he had leg pains again. When he was 32.5 years old, Theo had a persistent cough and Vincent described his brother's illness as "changeable and uneven." Theo (now 33.3 years old) was mobile and reasonable when he supervised Vincent's funeral at the end of July 1890. In August, Émile Bernard said that Theo "lost his reason, collapsed, was paralyzed." In September 1890, Theo complained of nightmares and hallucinations, but had the good sense to associate them with a new cough medicine. On October 12, 1890, he was suddenly hospitalized in Paris; he recovered sufficiently to be moved to Holland (under restraint) on November 17, 1890; he died just 15 weeks after his first hospitalization.

Theo may have contracted syphilis as a teenager. Bonger said that Theo was "stiff, as after a fall" in late 1886 at the age of 29. This may have been either paralysis (unable to move) or pain (not wanting to move). If they were paralytic attacks, then to have them at age 29 and die in the 34th year can be accommodated by a diagnosis of neurosyphilis. However, the fact that Theo had sustained periods of comparatively good health as well as intermittent problems in the intervening five years is entirely inconsistent with the unremitting and inexorably progressive nature of this disease. An extremely rare form of neurosyphilis involves tabetic and then paretic symptoms in the same individual. There have even been cases where the pain of tabes dorsalis resolves but it is most important to note that ataxia (failure of muscle coordination) and pupillary abnormalities do not. In fact, the spontaneous arrest of pain in such cases is actually due to nerve degeneration. In this state the patient has an unusually wide-based gait, exhibits foot-drop, and so on – signs that were never described for Theo. The course of his illness and the gamut of symptoms are much better explained by acute intermittent porphyria, and a fatal, porphyritic crisis explains the end.

Tralbaut in 1981 (reference 4, page 286) indicated that he was not only unconvinced about the diagnosis of neurosyphilis, but that he also realized the weight of evidence supporting an inherited disease. "Vincent's and Theo's mental and physical disabilities were similar in every respect. Moreover their younger sister Wilhelmien spent most of her life in establishments for neuropaths. There were therefore three members of the family who exhibited symptoms of varying degrees of mental disorder. The significance of this is inescapable. Even if we suppose that both brothers contracted venereal disease [Vincent in Antwerp and/or The Hague, and Theo in Paris], the probability of hereditary influence [as the underlying cause of their medical problems] seems to be overwhelming."

Ménière's disease and a case of misplaced surgery

In 1861, Prosper Ménière published several papers relating his observations on afflictions of the inner ear which caused nausea, vomiting, and vertigo. The disease was subsequently named after him and is characterized by hearing loss, vertigo, and tinnitis (ringing in the ears), and is usually unilateral (one ear only).[51] During an attack of vertigo the patient is completely orientated to his surroundings and has no neurologic deficit such as paresthesia, diplopia, loss of consciousness, weakness, or paralysis. Sounds are distorted in the affected ear and are perceived as "tinny." Loud sounds are intolerable or even painful, and hearing acuity gradually declines.

The initial episode of sudden-onset severe vertigo, accompanied by nausea and vomiting, usually lasts two to four hours. In the intervening period the patient may notice a fluctuation in hearing and roaring tinnitus. The frequency of these episodes varies from once a month through once or twice a week to the extreme of every day. Spontaneous remissions can occur with no residual symptoms other than some hearing loss in the involved ear.[52] Ménière's seems to be an acquired disease but the etiology is still uncertain.[51,52]

In 1979, Yasuda[53] wondered in print, "Was van Gogh suffering from Ménière's disease?" The twelve page article was published in Japanese, but contains a full two pages of introduction and summary in English, more than enough to grasp the author's thrust. Those speculations received little support then, because the diagnosis of Ménière's disease was based on a limited selection of symptoms. This dubious diagnosis was a sincere attempt, but it received little attention subsequently, except to be recorded in the most comprehensive bibliographies.

The *Journal of the American Medical Association* [JAMA], on the week of the centenary of Vincent van Gogh's death, declared that, "Van Gogh had Ménière's disease and not epilepsy."[54] It was wrong on both counts; there is no case for Ménière's disease and epilepsy was no longer even the diagnosis of merit. A Colorado ear specialist and his colleagues had rediscovered Yasuda's hypothesis and rewrote it as a definitive diagnosis. Their conclusion was based on a limited selection of symptoms, the pretense that epilepsy was the only viable alternative, and their propensity for construing certain complaints as hallmarks of the ear disease. Thus van Gogh's gastrointestinal problems were taken to be strictly nausea and vomiting, several references to hearing voices were relegated to tinnitus, and the psychosis that was grave enough to cause self-mutilation and eventual suicide was underplayed. Their claim that van Gogh severed the lower half of his left ear to relieve tinnitus must surely strike readers, if not the editors of JAMA, as misplaced surgery. It becomes an absurd suggestion when put into context; Vincent had no recollection of the incident.

Monroe[55] pointed out that Vincent's relapses after visiting Arles were more likely due to absinthe drinking than the motion sickness invoked by Arenberg et al.,[54] who were also criticised for misreading and misquoting several of van Gogh's letters. Furthermore, Vincent's lengthy train trip from St. Rémy to Paris, shortly thereafter, was notably uneventful and not indicative of a susceptibility to motion sickness. Thus the Ménière's disease hypothesis is one of the least attractive. The presentation[54] exemplifies all of the worst aspects of selective embrace of symptoms, and misconstruing of quotations in order to shore up an idea.

A reasonable hypothesis: intoxication with metal salts

> *If you know that it is dangerous for you to have colors near you, why don't you clear them away for a time, and make drawings?*
> Theo to Vincent, letter T23, from Paris, January 3, 1890.

In chapter 2, I discussed the evidence for Vincent van Gogh nibbling at his paints. The documentation of this bizarre behavior involves two very explicit letters from Theo (letters T23, T24) and an entry by Dr. Peyron in the register of the St. Rémy asylum. In the same context (letter 601, originally published in abridged form) Vincent discussed the accusation by others that he ate "dirt." Also, Tralbaut confirmed Vincent's attempts to eat paint by consulting with one of the asylum's attendants, albeit many years later. The available evidence indicates that this was not a conscious act on Vincent's part, and suggestions by others that it was an attempt at suicide have no credibility.

I proposed that Vincent had developed a pica for terpenes and it was the turpentine mixed with the pigments that tickled his taste-buds.[56] (Vincent's attempt to drink turpentine in Arles suggests that the habit had started earlier.) The chemical connection with Vincent's reckless use of camphor and the constituents of absinthe was made in chapter 4. Nonetheless, eating oil paints can hardly be recommended and it is worth looking at the chemistry and potential toxicology involved.

Artist's oil colors

Paint is a dispersion of a pigment in oil. The pigment is diluted and thinned with a solvent such as turpentine. When applied to a surface the volatile thinner evaporates leaving a film of pigment in the drying oil which slowly hardens. The so-called drying oils are glycerides with a high proportion of unsaturated fatty acids and the drying process is a combination of oxidation and polymerization.[57] The vehicle of the fine artist is almost exclusively linseed oil from flax seeds, *Linum usitatissimum.* The composition varies with the source but is always rich in linolenic (35–67%), linoleic (8–29%), and oleic (12–30%) acids.[58] When it is exposed to air there is an absorption of oxygen which is at first slow, for a time more rapid, and then slow again as the process nears completion. At ordinary temperatures the slow period of induction takes one to three days and the drying process is complete in about twenty to thirty days. The chemical composition of the pigment sometimes modifies the drying time.

The Egyptians as far back as 8,000 B.C. used mineral pigments of mercury, arsenic, copper, and iron, as well as some plant extracts. From the time of the Roman Empire through the Renaissance new pigments of lead, antimony, tin, aluminum and manganese were developed. In the nineteenth century further examples of heavy metal salts involving cobalt, chromium, and cadmium were introduced. It was not until the second half of the nineteenth century that coal-tar dyes and analine dyestuffs made a significant contribution to artist's pigments.[59]

The fabrication of artist's oil colors involves grinding the inorganic or organic pigments in linseed oil to a buttery paste. Present day toners contain about 30% pigment, but the dense zinc and lead whites, are as high as 80%. Modern methods involve roller mills but in Vincent's time purveyors of pigments prepared their products by hand. The range in quality depended on the degree of grinding as well as the raw pigments employed.[60]

Several of the mineral pigments used by Vincent van Gogh and his contemporaries contained salts which were relatively toxic and we shall consider mercury, chromium, and lead. Examples of pigments used by Vincent, which contained these elements, are vermilion or cinnabar, HgS, (letters 475, 584, 604); chrome yellow, $PbCrO_4$, (letters 475, 541, 541a, 551, 584, 608, 629); white lead or flake white, $2PbCO_3.Pb(OH)_2$, (letters 475, 541, 551, 584, 594); and red or orange lead, Pb_3O_4, (letters 475, 581, 584, 592, 629).

Mercury

Elemental mercury, organic compounds such as methymercury, CH_3Hg^+ and inorganic salts such as mercuric chloride, $HgCl_2$, can all present potential hazards.[61] With respect to the artist we should concern ourselves only with inorganic salts, and note that gastrointestinal absorption is usually of the order of 10% of the dose. These compounds dissociate into the mercuric ion, Hg^{2+} which becomes concentrated in the kidney, but does not cross the blood-brain barrier to a significant extent.

The acute effects of inorganic mercury poisoning include irritation to mouth and throat, nausea and vomiting, abdominal pain, weakness and fatigue. Kidney failure results after relatively high levels of ingestion. While organic compounds of mercury often produce a toxic neuroencephalopathy this is rarely seen with inorganic intoxication. The historically notable example of chronic inorganic mercury poisoning relates to the use of mercuric nitrate, $Hg(NO_3)_2$, in the felting process, producing the "mad hatters" of *Alice's Adventures in Wonderland* fame. It results in involuntary tremor and psychological abnormalities although the latter do not usually progress beyond nervousness, irritability, and pathological shyness. The primary salt that Vincent was exposed to was mercuric sulfide, in the pigment cinnabar (vermilion), which is practically insoluble in water. Neither the degree of exposure nor the course of his illness provides much support for a specific involvement of Vincent van Gogh with mercury intoxication.

Chromium and arsenic

Ingestion of large doses of chromium salts causes vertigo, abdominal pain and vomiting and may progress to convulsions, coma, and death.[61] The chronic effects are associated with ulceration, dermatitis and edema. Notwithstanding the toxicity of chromium salts, there seems to be little connection with the nature or the time course of Vincent's illness. Acute or chronic poisoning with arsenic can present with severe abdominal pain. However, I have not found evidence of arsenic salts, e.g. the orpiments, arsenic disulfide (red) or arsenic trisulfide (yellow), among the pigments mentioned in Vincent's letters. Likewise, there is no mention of cadmium yellows. The toxicity of these and other pigments should not be underestimated, but they do not appear to have been involved with Vincent. The most likely involvement centers on lead.

Lead

About one-third of patients with excessive exposure to lead suffer colicky, abdominal pain; fatigue, joint pains, headache, and irritability are also quite common. Impotence, constipation, vomiting, and diarrhea have all been observed to some extent. Subtle effects on personality, memory, and learning ability are frequently associated with chronic lead poisoning. However, seizures and confusional states are less common, especially in adults.[61,62]

Probably less than 10% of an ingested lead dose is retained, the remainder appears in the feces. On the other hand, as much as 70% of an inhaled dose is absorbed, provided the particle size is small enough to reach the alveoli of the lungs. The circulating blood and the soft tissues become the active pool of lead; this represents only about 10% of the total body burden. The other 90% appears in the bones where it is innocuous, but subject to reappearance in the blood. For this reason, a seemingly effective therapeutic treatment that depletes blood lead after chronic exposure often has to be repeated because the release of lead from the bones establishes a new toxic level in the soft tissues, and clinical symptoms recur.

Lead may be the oldest recognized chemical toxin; reports of occupational lead poisoning date to ancient Greece, and toxic levels have been found in Egyptian mummies. Artisans of lead-glazed pottery and stained glass were particularly susceptible to intoxication until better conditions were adopted in the workplace. The ingestion of paints containing lead pigments has, even up until recent times, presented a serious health hazard for children. Artists and craftsmen were exposed in the past because of their habit of wetting brushes orally and their accidental ingestion of lead-containing pigments from their tools and hands. The following case has been abstracted from a published report[63] and demonstrates a typical presentation as well as the dynamics of lead poisoning.

A 43 year old woman, employed as an art curator, was discovered during a routine medical examination in December 1978, to have a slightly lowered hemoglobin level in her blood (anemia). Two months later she complained of dizziness, diffuse muscle pains, and crampy abdominal discomfort unrelated to eating. She admitted to increasing irritability, and being unable to cope with her adolescent children. In mid-March of 1979, her hemoglobin level had fallen further and biochemical tests were initiated. Eight days later abdominal pain was worse. The concentration of lead in the blood was 72 µg / dL. (The currently accepted action level is 40 µg / dL.) A presumptive diagnosis of lead poisoning was made and the patient was asked to discontinue her restoration work as an art curator.

After three weeks of bed rest her symptoms were even worse and the hemoglobin level was now about one-half of normal. She was hospitalized and received an infusion of red blood cells. The lead concentration in her blood was now 127 µg/dL and she was excreting greatly elevated levels of δ-aminolevulinic acid and porphyrins in her urine. A course of chelation therapy was started over the next five days. (This involved ethylenediaminetetraacetic acid which has a high affinity for lead ions; it was administered as a mixture of calcium and sodium salts.) Ten days later the blood level of lead was 74 µg/dL and the patient reported less abdominal and muscle pain, and less irritability. Three weeks later the lead concentration was up to 84 µg/dL (even though she had not returned to work) and was accompanied by malaise and weakness in the legs. A second course of chelation therapy was instituted and her condition improved steadily. By September she was back to normal in terms of hemoglobin level, the lead concentration was 27 µg/dL, and all the symptoms of lead poisoning had disappeared.

Comments on the case of lead poisoning

The art curator had been restoring a Peruvian tapestry from about 1250, using a red pigment, cinnabar, from the same period. It was applied as a suspension in water. After drying, the fabric was extensively brushed and this produced considerable dust in the poorly ventilated room. The patient inhaled considerable quantities. She also repaired the tapestry and admitted to frequently wetting the sewing thread in her mouth.

Cinnabar (vermilion) is a mineral pigment composed almost exclusively of mercuric sulfide, HgS, which was originally mined, but later manufactured from mercury and sulfur. (Both types have the same chemical composition although the older, native mineral pigments sometimes had a different distribution of broken crystal particles than the more modern preparations.) The particular pigment employed by the patient was found to contain 1% lead oxide (red lead), Pb_3O_4, supposedly added as a fabric preservative.

Mercuric sulfide is very insoluble and the curator's exposure apparently did not involve substantial mercury absorption as indicated by normal urine concentrations. The lead content was sufficient to cause intoxication. This case also exemplifies the rebound phenomenon which is typical with accumulated lead. The partial clearing of lead by the first course of chelation therapy was followed by an increase, which was due to release from the bones. This was satisfactorily rectified by the second course of chelation.

Biochemistry of lead poisoning

Lead has an affinity for functional sulfhydryl groups in enzymes generally and a particularly sensitive example is δ-aminolevulinic acid dehydratase. This is enzyme number two in the heme biosynthetic pathway (Figure 6.2) and the inhibition accounts for excessive excretion of δ-aminolevulinic acid (ALA) in the urine of lead-intoxicated patients (see the case above). The last enzyme in the pathway, ferrochelatase (Figure 6.2), which catalyzes the incorporation of iron into protoporphyrin to form heme, is also inhibited by lead and this also contributes to the observed anemia.[61]

The excessive production of ALA in lead poisoning is similar to that found in acute intermittent porphyria, but note that porphobilinogen does not accumulate in lead poisoning. The similarity in neurological symptoms between AIP and lead poisoning may be in part referable to ALA. In experimental models it has been shown that lead chloride, $PbCl_2$, induces a modest rise in ALA synthetase activity as do the porphyrinogenic drugs such as phenobarbital. Most interesting is the observation that the induc-

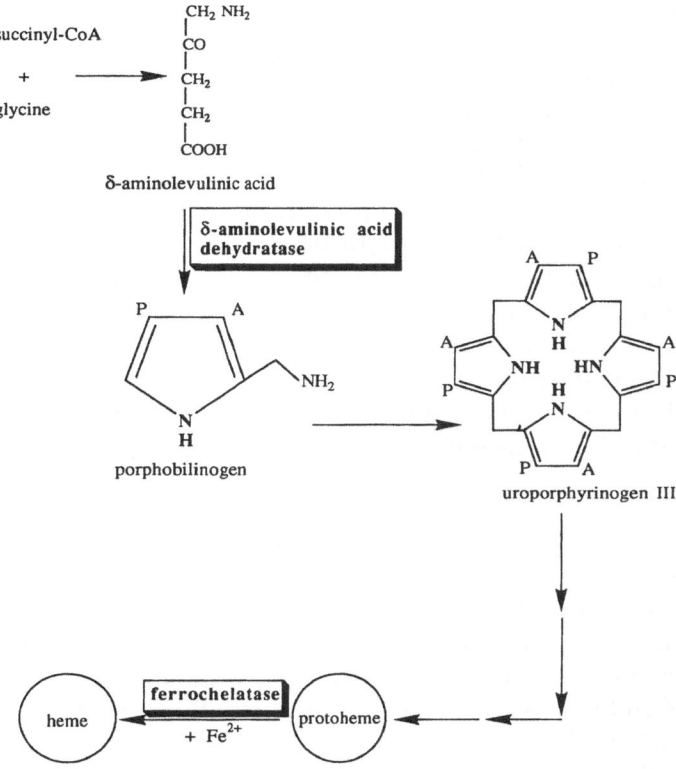

Figure 6.2. The heme pathway. Enzymes that are particularly sensitive to lead are indicated.

tion of this rate-limiting enzyme is greatly enhanced by adding lead chloride and phenobarbital together, i.e. a synergistic effect.[64]

Vincent and lead intoxication versus AIP

Abdominal pain, constipation, vomiting, paralysis, or paresis are very common in both AIP and lead poisoning. Neuropsychiatric symptoms are sometimes observed with lead intoxication but much less frequently than in acute intermittent porphyria.[65] There was no chelation therapy for lead poisoning in Vincent's time, and if his ingestion of lead salts (from his pigments) had been chronic, then the time course of such an illness would have been relentless and not episodic, as is well documented for van Gogh. The exposure to lead pigments may have been an added factor. In chapter

2, I suggested that the exceptional crisis in St. Rémy, the one that could not be related to drinking visits to Arles, might be connected with paint and turpentine nibbling.

Those who suggest a more active involvement for Vincent with lead or other metals seem to have restricted their ideas to comments in newspapers and magazines. I have also received a personal communication along this line from the owner of a van Gogh painting who feels that an embedded hair might be Vincent's and might be worthy of metal analysis. I pointed out that with time the hair has probably equilibrated with the pigments and will not be too indicative of the status of the donor. The present concern about environmental and workplace contamination has undoubtedly sharpened the interest in this type of intoxication. However, heavy metal intoxication will not explain the full story with Vincent. Furthermore, it is quite unlikely that Theo and Wil van Gogh were exposed to lead.

Miscellaneous psychiatric disorders

Tralbaut (reference 4) lists a score of further diagnoses and their authors. Many of them are variations upon the themes of epilepsy and schizophrenia, sometimes phrased in archaic or peculiar language. None advances the field or is worth discussing here. Perusal of the *Diagnostic and Statistical Manual of Mental Disorders,* (DSM), which is published by the American Psychiatric Association, provides a plethora of categories and subcategories that might describe various behavioral episodes or patterns in the life of Vincent van Gogh. Conduct and anxiety disorders, for example, offer useful words for some of Vincent's behavior but tell us nothing about his underlying illness.

Summary

I have discussed several hypotheses which have been put forward in the past or arise in the differential diagnosis. Although many can explain particulars, none offers a more reasonable alternative to acute intermittent porphyria (AIP).[1] With regard to the exacerbating factors, it is intriguing to recall the anecdotal diagnosis[2] attributed to Dr. Gachet, "an intoxication due to turpentine and also an intoxication from a sun too intense for a Nordic head." Notwithstanding some generosity toward the unwritten word of Paul Gachet, and perhaps some wishful thinking about his insight, I quote from a 1985 paper by Henry Peters, "exposure to oil-based paints and solvents will, in some porphyrics, produce symptomatology including psychosis, colic, seizures, and neuropathy. Very rarely in acute porphyria, extreme exposure to sunlight may provoke an attack."[66]

References

1. Loftus LS & Arnold WN. 1991. Vincent van Gogh's illness: acute intermittent porphyria? *British Medical Journal* **303**: 1589–1591.
2. Beer J. 1935. *Essai sur les Rapports de l'Art et de la Maladie de Vincent van Gogh*. Medical Thesis, University of Strasbourg.
3. Perry I. 1947. Vincent van Gogh's illness: a case record. *Bulletin of the History of Medicine* **21**: 146–172.
4. Tralbaut ME. 1981. *Vincent van Gogh*. New York: The Alpine Fine Arts Collection Ltd.
5. Lubin AJ. 1987. *Stranger on the Earth: A Psychological Biography of Vincent van Gogh*. New York: Henry Holt & Co.
6. Niedermeyer E. 1983. *Epilepsy Guide. Diagnosis and Treatment of Epileptic Seizure Disorders*. Baltimore & Munich: Urban & Schwarzenberg.
7. Commission on Classification and Terminology of the International League Against Epilepsy. 1981. Proposal for revised clinical and electroencephalographic classification of epileptic seizures. *Epilepsia* **22**: 489–501.
8. Penry JK & Rakel RE. 1986. *Epilepsy: Diagnosis, Management, Quality of Life*. New York: Raven Press.
9. Penry JK. 1975. Perspectives in complex partial seizures. chapter 1, in: *Advances in Neurology: Volume 11, Complex Partial Seizures and their Treatment*. JK Penry & DD Daly (eds). New York: Raven Press.
10. Newmark ME & Penry JK. 1980. *Genetics of Epilepsy: a Review*. New York: Raven Press.
11. Navratil L. 1959. Vincent van Gogh: his disease assessed in the light of his paintings. *CIBA Symposium* **7**: 210–216.
12. Schnier J. 1950. The blazing sun: a psychological approach to van Gogh. *American Imago* **7**: 143–162.
13. Jaspers K. 1922. *Strindberg und van Gogh*. Leipzig: Ernst Bircher.
14. Riese W. 1925. Über den Stilwandel bei van Gogh. *Zeitschrift für die gesamte Neurologie und Psychiatre* **98**: 1–16.
15. Minkowska F. 1933. Van Gogh: les relations entre sa vie, sa maladie, et son oeuvre. *L'Evolution Psychiatrique* **3**: 53–76.
16. Jackson JH. 1931. *Selected Writings of John Hughlings Jackson*. London: Hodder & Stoughton.
17. Gibbs FA, Gibbs EL & Lennox WG. 1937. Epilepsy: a paroxysmal cerebral dysrhythmia. *Brain* **60**: 377–388.
18. Penfield W & Jasper H. 1954. *Epilepsy and the Functional Anatomy of the Brain*. Boston: Little, Brown and Co.
19. Gastaut H. 1970. Clinical and electroencephalographical classification of epileptic seizures. *Epilepsia* **11**: 102–113.
20. Doiteau V & Leroy E. 1928. *La Folie de Vincent van Gogh*. Paris: Éditions Æsculape.
21. Vinchon J. 1960. Diagnostic de la "folie" de van Gogh. *Historie de la Médecine Communications présentées à Paris à la Société Francaise d'Histoire de la Médecine en 1960*. pp 23–24.

22. Godlewski G. 1982. Vincent van Gogh, prince des maudits. *Diamant Actualités Médicales* **29**: 12–16.
23. Pickvance R. 1986. *Van Gogh in Saint-Rémy and Auvers.* New York: Harry N. Abrams Inc.
24. Hemphill RE. 1961. The illness of Vincent van Gogh. *Proceedings of the Royal Society of Medicine* **54**: 1083–1088.
25. Monroe RR. 1978. The episodic psychoses of Vincent van Gogh. *The Journal of Nervous and Mental Diseases* **166**: 480–488.
26. Victor M & Adams RD. 1977. Alcohol. chapter 118, in: *Harrison's Principles of Internal Medicine.* 8th ed. GW Thorn et al. (eds.). New York: McGraw-Hill Book Co.
27. Berrios GE. 1988. Depressive and manic states during the nineteenth century. chapter 2, in: *Depression and Mania.* A Georgotas & R Cancro (eds.) New York: Elsevier.
28. Pinel P. 1818. *Nosographie Philosophique ou la Méthode de l'Analyse Appliquée a la Médicine.* 6th ed. Paris: Brosson.
29. Esquirol JE. 1820. Mélancholie. in: *Dictionnaire des Sciences Médicales par une Société de Médicins et la Chirurgienms.* Paris: Panckoucke.
30. Falret JP. 1854. Mémoire sur la folie circulaire. *Bulletin de l'Académie de Médicine* **19**: 382–415.
31. Baillarger JGF. 1854. De la folie à double-forme. *Annales Medico-Psychologiques* **6**: 367–391.
32. Kräpelin E. 1921. *Manic-depressive Insanity and Paranoia.* (translated by RM Barclay) 8th ed. Edinburgh: Livingstone.
33. Fabbri R Jr. 1966. Dr. Paul-Ferdinand Gachet: Vincent van Gogh's last physician. *Transactions and Studies of the College of Physicians of Philadelphia* **33**: 202–208.
34. Sarwer-Foner GJ. 1988. The course of manic-depressive (bipolar) illness. chapter 4, in: *Depression and Mania.* A Georgotas & R Cancro R (eds.) New York: Elsevier.
35. Sarwer-Foner GJ. 1966. A psychoanalytic note on a specific delusion of time in psychotic depression. *Canadian Psychiatric Association Journal* **11** (supplement): S221–S228.
36. Goodwin FK & Jamison KR. 1990. *Manic-depressive Illness.* New York: Oxford University Press.
37. Sussman N & Cancro R. 1988. Differential diagnosis of manic-depressive and schizophrenic illness. chapter 11, in: *Depression and Mania.* A Georgotas & R Cancro (eds.) New York: Elsevier.
38. Roose SP, Glassman AH, Walsh BT, Woodring S & Vital-Herne J. 1983. Depression, delusions, and suicide. *American Journal of Psychiatry* **140**: 1159–1162.
39. Andreasen NC & Glick ID. 1988. Bipolar affective disorder and creativity: implications and clinical management. *Comprehensive Psychiatry* **29**: 207–217.
40. Goodwin DW. 1988. *Alcohol and the Writer.* Kansas City: Andrews & McMeel.
41. Baldessarini RJ & Lipinski JF. 1977. The schizophrenia syndrome and related psychoses. chapter 346, in: *Harrison's Principles of Internal Medicine.* 8th ed. GW Thorn et al. (eds.). New York: McGraw-Hill Book Co.
42. Berndl K, von Cranach M & Grüsser OJ. 1986. Impairment of perception and recognition of faces, mimic expression, and gestures in schizophrenic patients. *European Archives of Psychiatry and Neurological Sciences* **235**: 282–291.

43. Grüsser JO, Selke T & Zynda B. 1988. Cerebral lateralization and some implications for art, aesthetic perception, and artistic creativity. chapter 11, in: *Beauty and the Brain.* I Rentschler, B Herzberger & D Epstein (eds.). Basel, Boston, Berlin: Birkhäuser Verlag.

44. Prinzhorn H. 1972. *Artistry of the Mentally Ill.* (translated by E von Brockdorff) New York, Heidelberg, Berlin: Springer-Verlag. (Original German version: Prinzhorn H. 1922. *Bildnerei der Geisteskranken.* Berlin: Verlag Julius Springer.)

45. Grey R. 1924. *Vincent van Gogh.* Rome: Valori Plastici.

46. Beer J. 1950. Van Gogh: diagnosis of the tragedy. *Art News Annual* **19**, special number: 82–90.

47. Fabricant J. 1958. Heat stroke. *U.S. Armed Forces Medical Journal* **9**: 1106–1118.

48. Satterthwaite TE. 1886. Heat-stroke. vol 3, p 600, in: *A Reference Handbook of the Medical Sciences.* AH Buck (ed.). New York: William Wood.

49. Robbins SL. 1957. *Textbook of Pathology with Clinical Applications.* Philadelphia & London: WB Saunders Co.

50. Goodman LJ & Karakusis PH. 1988. Neurosyphilis. vol 52, chapter 18, in: *Handbook of Clinical Neurology.* PJ Vinken, GW Bruyn & HL Klawans (eds.) Amsterdam: Elsevier Science Publishers.

51. Harker LA & McCabe BF. 1980. Meniere's disease and other peripheral labyrinthine disorders. chapter 41, in: *Otolaryngology.* 2nd ed. MM Paparella & DA Shumrick (eds) Philadelphia: WB Saunders.

52. Ballenger JJ. 1984. *Diseases of the Nose, Throat, Ear, Head, and Neck.* 13th ed. Philadelphia: Lea & Febiger.

53. Yasuda K. 1979. Was van Gogh suffering from Ménière's disease? *Otologia Fukuoka* **25**: 1427–1439.

54. Arenberg IK, Countryman LF, Bernstein LH & Shambaugh GE. 1990. Van Gogh had Ménière's disease and not epilepsy. *Journal of the American Medical Association* **264**: 491–493.

55. Monroe RR. 1991. Another diagnosis for Vincent van Gogh? *The Journal of Nervous and Mental Diseases* **179**: 241.

56. Arnold WN. 1988. Vincent van Gogh and the thujone connection. *Journal of the American Medical Association* **260**: 3042–3044.

57. Gunstone FD. 1958. *An Introduction to the Chemistry of Fats and Fatty Acids.* New York: John Wiley & Sons Inc.

58. Kirschenbauer HG. 1960. *Fats and Oils: an Outline of their Chemistry and Technology.* New York: Reinhold Publishing Corporation.

59. Patton TC. 1973. *Pigment Handbook.* New York: John Wiley & Sons.

60. Gettens RJ & Stout GL. 1966. *Painting Materials: A Short Encyclopaedia.* New York: Dover Publications Inc.

61. Ellenhorn MJ & Barceloux DG. 1988. *Medical Toxicology.* New York: Elsevier Science Publishing Company Inc.

62. Dagg JH, Goldberg A, Lochhead A & Smith JA. 1965. The relationship of lead poisoning to acute intermittent porphyria. *Quarterly Journal of Medicine* **34**: 163–175.

63. Fischbein A, Wallace J, Anderson KE, Sassa S, Kon S, Rohl AN & Kappas A. 1982. Lead poisoning in an art conservator. *Journal of the American Medical Association* **247**: 2007–2009.

64. Maxwell JD & Meyer UA. 1978. Pharmacogenetics in the field of heme metabolism: drug sensitivity in hereditary hepatic porphyria. chapter 8, in: *Heme and Hemoproteins.* F De Matteis & WN Aldridge (eds.) Berlin: Springer-Verlag.
65. Sassa S. 1978. Toxic effects of lead, with particular reference to porphyrin and heme metabolism. chapter 11, in: *Heme and Hemoproteins.* F De Matteis & WN Aldridge (eds.) Berlin: Springer-Verlag.
66. Peters HA. 1986. Acute hepatic porphyria. pp 317–321, in: *Current Therapy in Neurologic Disease 1985–1986.* RT Johnson (ed.) New York: BC Decker.

Chapter 7
Vincent's Doctors

Figure 7.1. Vincent van Gogh: *Portrait of Dr. Gachet*, 1890, Auvers-sur-Oise, oil on canvas, 68 × 57 cm (26.8 × 22.4″), Paris, Musée d'Orsay, © Photo R.M.N.

*Go and consult a doctor every day, not
particularly because a doctor could do
something for you that would prove effective,
but because the measure of going to see a doctor,
etc., would force you to bear in mind, this is
nervousness, this is what I am. It is a question
of self-knowledge, of serenity notwithstanding
all the tricks nerves must play.*
Vincent to Theo, letter 332, from
Drenthe, Fall, 1883.

The deep-seated problems that beset Vincent van Gogh were not reversed, and his suicide labels the case a medical failure by any standard. However, the events should be placed in the context of nineteenth century medicine, and this may at least allow us to accommodate the sad outcome of the artist's illness. Brief biographies of some of the doctors who touched van Gogh, either directly or indirectly, will help to round out the picture.

It has been tacitly assumed in some quarters that a modern psychiatrist, armed with all manner of miracle drugs, could step in tomorrow and achieve a wonderful rehabilitation of a latter day Vincent van Gogh. Given the hypothesis developed in chapter 5, this bright scenario would not follow inevitably. The successful management of acute intermittent porphyria depends primarily on good nutrition and the avoidance of precipitant factors such as alcohol and, as shown by our own studies, the terpenes of absinthe. Some of the favorite drugs of twentieth century medicine, if applied in our hypothetical case, would soon be discovered by trial and error to do nothing but exacerbate the crises of the porphyric.

In order to avoid medical crises these individuals should be well nourished and lead a careful and balanced life style. They would be advised to avoid periods of fasting, not to abuse alcohol, and not to smoke. In my opinion, the saddest aspect of Vincent's illness was that some of his own physicians gave him this very advice, but that the artist did not comply.

Nineteenth century French psychiatry[1,2]

It is appropriate to start with the evolution of the discipline of psychiatry, as we would call it today, for two reasons. First, practitioners with a bent towards treating the mentally ill were obviously involved in offering Vincent nursing care, advice, and what little drug therapy was available at the time. And second, it is instructive to trace the maturation of ideas and concepts toward the insane and asylums. Social and political overtones were always present and are still important today.

The era starts with Philippe Pinel (1745–1826), an internist who took a special interest in mental illness. His famous unchaining of the insane at the Bicêtre in 1793 and the Salpêtrière in 1795 were landmarks in humane treatment of the mentally ill. He developed the concept of the therapeutic institution, proposed separately housed categories of inmates, rejected bleeding and purging, and encouraged a firm but benevolent attitude towards all patients. His *Medico-philosophical Treatise on Mental Alienation* (1801) emphasized clinical observation, statistical evaluation, and the rehabilitative power of work therapy. His most important novel category was *manie sans délire* (mania without delirium) or *folie raisonnante* (rational madness), to describe a form of insanity in which the intellectual functions appeared to remain intact.

Jean Dominique Esquirol (1772–1840) was a full-time mental health specialist, and possibly the first to live with his patients within the institutions he attended. In 1799, he aligned himself both philosophically and politically with Pinel, and twelve years later became chief physician at the Salpêtrière. In 1826 he moved to Charenton. He ran private institutions concurrently, and gave the first formal course in psychiatry in Paris, which soon attracted an international audience.

Esquirol was also one of the writers of the famous law of 1838, which is still in place. Accordingly, each French Département was required to create a lunatic asylum which would accommodate either voluntary or official admissions. The former, *placement volontaire*, was decided by the patient's family or close relatives, who could also obtain a discharge at any time. This was the format followed for Vincent's internment at St. Rémy. The official admission, *placement d'office*, was decided by the *préfet* (chief administrator of a department or province), responding to emergency or danger. In either case, several physicians were supposed to attest to the patient's condition, and if the patient protested against the confinement, a new investigation was supposed to be ordered. A similar situation prevailed in Holland in the 1880's. Vincent referred to this from The Hague (letter 206) in connection with his father's earlier threat to have him confined at Gheel. It will be recalled that Vincent hoped to collect support from his own doctors against the possibility of involuntary confinement.

Esquirol had a special interest in hallucinations but his approach was eclectic. He has been judged a better statistician, observer, and analyst than Pinel. In the field of systematics he propagated Pinel's *folie raisonnante* under the term *monomanie affective* (monomania). His pupils included Falret and Baillarger.

Jean Pierre Falret (1794–1870) founded an organization for the protection of discharged patients and practiced "mutual instruction" of the sick. His description, in 1854, of patients with alternate moods of mania and melancholia, *folie circulaire* (circular madness), was a major contribution for which he shared priority with Jules Baillarger (1809–1890). Baillarger wrote extensively on "double-formed" mental disease; his emphasis on the intervening period of normalcy made his concept somewhat closer than Falret's to modern descriptions of manic depressive psychosis. He was also instrumental in founding a specialty journal, *Annales Médico-Psychologiques,* in 1843, and a professional society, *Société Médico-Psychologique,* in 1847.

Possibly the most outstanding psychiatric and neurologic advance of the nineteenth century was made by Antoine Laurent Bayle (1799–1858) who described, in 1822, *dementia paralytica* (a form of neurosyphilis) as a progressive psychosis from monomania through mania to dementia. According to Ackerknecht, this marked the end of ancient psychiatry because it became clear that the manias and melancholias of the past were symptoms and not diseases. Bayle's insight was not properly rewarded, and he fell victim to politics in 1825 when he lost his position at Charenton to Esquirol. Bayle was forced to spend the remainder of his career as a librarian and an editor.

In the second half of the nineteenth century French psychiatrists became preoccupied with *degeneration* theory, which was inspired by new discoveries in evolution and heredity. Benedict Morel (1809–1873) searched for the causes of mental retardation in infantile illnesses and hereditary constitution. Accordingly, problems as diverse as misery, infection, and alcoholism were lumped under one roof, invoked as the primary cause of deviation from the norm, which was deemed to be inheritable (sometimes skipping generations, i.e. atavism) and then labelled degeneration. Valentin Magnan (1835–1912), who made so many contributions in connection with alcoholism and absinthism (see chapter 4), embraced Morel's views as did Krafft-Ebing (1840–1903) in Germany. Together with Jacques Moreau de Tours (1804–1884), Ernest Lasègue (1816–1883), and Jules Falret (1824–1902), they postulated that anxiety leads through delirium and hallucinations to megalomania and ultimate dementia. Moreover, this progression was conceived as virtually inevitable. This attempt at unification had the charm of simplicity and more or less excused the physician from intervention. It was too simplistic and obviously did not stand the test of time. Nonetheless, the concept of degeneration was remarkably influential at the turn of the century. Vincent and Theo van Gogh conversed along these lines on several occasions.

From the 1820's through the 1850's American medical men flocked to Paris, which had replaced London and Edinburgh as the medical Meccas.[3] All of the evolving specialties were well represented in France during those years. However, during the second half of that century, and the turn into ours, the interest shifted from France to Germany and England. This resulted from the relative expertise in chemistry, the rise in cytological pathology as a discipline, novel pursuits in drug therapy, and national competition.

A note on the doctoral thesis

For the nineteenth century physician, formal training after the baccalaureate involved practical training as an intern and a series of externships at teaching hospitals, the individual programs being identified as much by the director of the department as by the institution. This course of practical and theoretical exposure lasted six or more years, was typically performed at several hospitals, and progressed to a climacteric in which the candidate submitted, and then defended, a thesis to a faculty of medicine at a university. The chosen university was sometimes in a different city; for example, Dr. Gachet completed several externships in Paris but took his degree from Montpellier.

The medical thesis varied greatly in scope, substance, and scientific merit. Few candidates spent more than a year on their projects. The majority of theses were evaluations of the literature spiced with personal experiences and, judging from the titles, often assigned in areas far from the candidate's previously demonstrated interests in externships, or eventual practice. I hasten to add that this was not always the case, one example that springs to mind is the outstanding 1882 thesis of Philippe Gaucher (1854–1918) on a glycolipid storage disease that is now named for him. But this is the exception rather than the rule, and the thesis title was not always indicative of the doctor's future contributions or interests. For the most part we should look to the externships for specialty training, and these are usually listed and dated among the first few pages of the thesis, along with historically valuable lists of mentors.

The medical schools of France, Germany, and several other European countries, still require an M.D. thesis of the graduating physician; they still vary in depth and merit. None of the above is to be confused with thesis requirements for the doctor of philosophy degree in the sciences such as chemistry, wherein the formulation of a research problem, planning and execution of experiments, evaluation of data, and presentation within the framework of the extant literature, are the major activity for the Ph.D. candidate in the three to six years after completion of the bachelor's degree.

Vincent's doctors in Paris and Auvers

Louis Rivet

> *After all doesn't it do one all the good in the world to listen to the wise advice of Rivet and Pangloss [see Voltaire's Candide], those excellent optimists of the pure and jovial Gallic race, who leave you your self-respect?*
> Vincent to Theo, letter 481, from Arles, May 4, 1888.

Both Vincent and Theo consulted Dr. Rivet, and they recommended him to their friend Arnold Konig (letter 464). Rivet was born in 1851 at Saint Jean de Luz[4] (Basses Pyrénées, near the Spanish border), became an intern in the hospitals of Paris, and subsequently maintained a private practice at 6 rue de la Victoire for the bulk of his career. He wrote a 47 page thesis on pleurisy in infants which he defended before the Paris faculty in 1880. He also received formal training in psychiatry, including a stint at the Salpétrière in 1877, under Jacques Joseph Moreau de Tours (1804–1884), pupil of Esquirol and one of the authors of the degeneration hypothesis.

Another famous mentor was Alexandre Brierre de Boismont (1797–1881) whose arguments effectively buried the legend of the menstrual origin of insanity in women. Brierre de Boismont collaborated in the establishment of *Annales Médico-Psychologiques* in 1843, and he shared editorial duties with Baillarger and Cerise from 1850 to 1855. He was widely published and his books included three editions of *Des Hallucinations* (1845, 1852, 1861) and two editions of *Du Suicide et la Folie Suicide* (1856, 1865). Thus Dr. Rivet had both experience and contacts within psychiatry. His initial practice was at 55 rue du Faubourg-Montmartre, close to Theo van Gogh's apartment.

When Vincent started thinking about his illness, a few months after arrival in Arles, he came up with the following analysis. "My poor boy, our neurosis, etc., comes, it's true, from our way of living, which is too purely the artist's life, but it is also a fatal inheritance, since in civilization the weakness increases from generation to generation. If we want to face the real truth about our constitution, we must acknowledge that we belong to the number of those who suffer from a neurosis which already has its roots in the past" (letter 481). Apart from showing great insight into the inherited nature of his own and Theo's illness, the philosophy displayed in this letter is almost directly based on the then current theory of degeneracy, something that Rivet probably espoused from Moreau de Tours.

It would appear that the brothers respected Rivet and yet they weighed his treatments against the very different approaches of Dr. David Gruby. For example, when Rivet had Theo on a course of potassium iodide (a widely prescribed drug at the time but now thought to be rather ineffectual except for certain systemic fungal diseases), Vincent advised, " I should have it out with Rivet if Gruby tells you not to take any. I am sure that in any case you mean to keep on being friends with both ... Anyway you must have it out frankly with Rivet; he ought not to be jealous" (letter 489). They seemed to favor Gruby for the long term but Rivet for the reality of their habits, "Rivet takes things as they are, without trying to create a paradise, and without in any way trying to make us perfect. But he forges a cuirass [defensive armor for the body], or rather he hardens one against illness, and keeps up one's morale, I do believe, by making light of the disease one has got " (letter 489). It was Dr. Rivet who finally achieved a successful treatment for Theo's cough in 1889 (letters 602, T13, T14).

The potential role for Dr. Rivet, upon Vincent's return to Paris and Auvers, has been understated in the literature. In September 1889, Theo cautioned Vincent, "According to Rivet, and also according to what I conclude from M. Peyron's letters, there is a possibility which you would rather not believe, but it is absolutely necessary that you do nothing imprudent, and that you stay under the supervision of a doctor" (letter T17). This indicates that Theo had already been in consultation with Rivet about Vincent. His brother seemed to concur: "Now, if ... I should make a move then let's begin as if practically nothing was wrong, being very cautious all the same, and ready to listen to Rivet in the smallest matters" (letter 607). The arrangement with Dr. Gachet superseded and during Vincent's brief respite in Paris, and the two months in Auvers, there was no evidence of his consulting Rivet. This was indeed unfortunate because a professional comparison with Vincent's condition of three years earlier might have been instructive. A letter[5] from Theo's brother-in-law, Andries Bonger, in late 1890, relates an opinion on Theo's final illness: "Rivet said that his case is far worse than Vincent's, and that there is not a spark of hope." The basis for this is not clear; did Theo consult Dr. Rivet at this time or was the comparison based on earlier information?

David Gruby[6,7]

> *I think Gruby is right about [our] cases – to eat*
> *well, to live well, to see little of women, in short*
> *to arrange one's life in advance exactly as if*
> *one were already suffering from a disease of the*
> *brain and spine, without counting the neurosis*
> *which is actually there.*
> Vincent to Theo, letter 481, from Arles,
> May 4, 1888.

Vincent consulted Dr. Gruby while he was in Paris, and Theo continued to consult him, apparently as a second opinion to Dr. Rivet. He is mentioned in eight letters. Vincent was not optimistic about Gruby being around "when the time comes when we shall need him most" (letter 490), but the doctor survived them by eight years. Gruby stressed personal hygiene and sensible habits over the long term, "I am very curious to see the result [in you, Theo] of Gruby's system – in the long run – say, after a year's trial. It would be wise to go and show yourself to him sometimes, and chat with him, and really to catch his attention, a really serious effort on his part, just as B.[Bernard?] at last managed to get his sympathy and a more serious interest" (letter 492).

Gruby was born in Kis-Kér, Hungary, in 1810. He completed early studies at Pest (one of the twin cities of Budapest) and then finished his medical degree at Vienna, in 1839. He built his own microscope, and other instruments, and became very proficient in pathologic examination. He published his doctoral dissertation on leucocytes, as the first part of a promised treatise which was never completed. He published about 30 scientific papers, two-thirds of which were in volumes 13–34 of *Comptes rendus hebdomadairess des séances de l'Académie des sciences.*

He migrated to Paris and communicated his expertise to several students who later made contributions, not least of which Claude Bernard (1813–1878). In the 1840's Gruby published on several aspects of human mycology including the ringworm fungi. The importance of his findings were not fully appreciated until some 50 years later. In 1847–48, Gruby also performed experiments with ether and chloroform for general anesthesia. With regard to academic contributions there is no doubt that Dr. Gruby was by far the most accomplished of the medical men who saw Vincent van Gogh.

Perhaps because of disappointment about the meagre scientific recognition he received initially, Dr. Gruby subsequently devoted himself to general medicine and built up a large and notable practice in Paris, at 66 rue Saint-Lazare. Frédéric Chopin, Franz Liszt, Alexandre Dumas *père et fils,* Heinrich Heine, Alphonse Lamartine, Alphonse Daudet, George Sand, and the van Gogh brothers are counted among his distinguished patients. Dr.

Gruby wrote long prescriptions consisting mostly of water and harmless additives while offering reasonable advice on the troubles of everyday life. His most remarkable achievements were with hypochondriacs who were given seemingly convoluted regimens, elaborate directions, and strict instructions. For example, he prescribed for the neurasthenia of Alexandre Dumas père, "Take your early promenade at six in the morning; purchase three apples from the grocer; eat the first apple at the Arc de Triomphe, the second at the quai d'Orsay, the third at la Place de la Madeleine; after that you may go to your house, but always on foot; do this for fourteen days and then visit me again." Another patient was enjoined to eat gingerbread and cheese at five in the morning exactly, and then apples at midnight. Meticulous execution by the patient often resulted in cure, apparently by diversion, but perhaps also by indirect substitution of more exercise and better nutrition.

Gruby was rather restricting in his recommendations on sex. Nevertheless, both Vincent and Theo were prone to invoke Gruby's maxims: "The desire for women that you catch in Paris, isn't it rather the effect of that very enervation of which Gruby is the sworn enemy, rather than a sign of vigor? … The root of the evil lies in the constitution itself, in the fatal weakening of families from generation to generation, and besides that, in one's unwholesome job and the dreary life of Paris" (letter 521). Again, we see the preoccupation with degeneration theory.

Dr. Gruby was a quiet and singular person who lived alone. His personal habits, for example sleeping on the floor with only animal skins, added to his reputation as an eccentric.[8] None of that should detract from his lasting contributions in basic science and his successes in treating psychosomatic medical problems. He remained in Paris where he died in 1898, and is buried at Saint Vincent Cemetery in Montmartre.

Paul Ferdinand Gachet

> *With the greatest regret I must disturb your repose. Yet I think it is my duty to write to you immediately. At nine o'clock in the evening of today, Sunday, I was sent for by your brother Vincent, who wanted to see me at once. I went there and found him very ill. He has wounded himself.*
> Dr. Paul Gachet to Theo, from Auvers-sur-Oise, July 27, 1890.

Paul-Ferdinand Gachet[9,10] was not an ordinary fellow. He was born to a textile mill owner in 1828 in the northern French city of Lille. He spent most of his adult life in Paris and Auvers-sur-Oise. He became a physician and his patients included a lady from as far away as the Virgin Islands; a middle-aged man whom he advised against a leg amputation; some of the first children to receive powdered milk as a dietary supplement; and several psychiatric cases. Such diversity was not exceptional for nineteenth century France; however, the old lady from the West Indies was Camille Pissarro's mother, Rachael, and Gachet went on to attend three generations of that artistic family; the amputee was the painter Édouard Manet who refused the advice and died eleven days later; Gachet's chemist friend who introduced powdered milk into France was Nestlé (before he became a name brand); and his psychiatric cases included the etcher Charles Méryon. For these and other encounters the Doctor's life was full and extraordinary.

Dr. Gachet's training was admirable. He completed the requirements for the bachelor's degree at the University of Paris, and then loaded himself with experiences in the mental hospitals (Bicêtre and Salpêtrière) where he worked in the services of J.P. Falret and J. Baillarger. He counted among his teachers the great Armand Trousseau (1801–1867), last of the classical clinicians according to Ackerknecht.[1]

Gachet was awarded the medical degree, in 1858, for a thesis[*] on melancholia at Montpellier Medical School, a venerable establishment whose longevity for instruction is second only to Bologna. Gachet's thesis, which is mostly a literature review with some evaluations, starts with a philosophical statement, mentions the great men of science and letters who were afflicted with melancholia, and goes on to speculate that whole societies and nations may have been affected at various times. In the clinical arena he allowed the use of warm baths for calming anxious patients but condemned blood letting and purging (still prevalent in mid-nineteenth century). The dissertation is well written and free of trite clichés.

After returning to Paris, Gachet set up private practice; he also donated services at some of the dispensaries. In the next few years he was impressed by Vincent Simon who espoused homeopathy, an approach made popular by Samuel Hahnemann (1755–1843), a German physician who spent the bulk of his later career in Paris. The homeopathic approach had its origins in much earlier work which included the experiments of Robert Boyle (1627–1691). It was based on "like cures like," and proposed essentially that if a plant extract produces in a healthy animal a symptom resembling that of a specific illness, then small doses of the same should be given to the

---

[*] *Étude sur la Mélancolie*, 111 pages, Montpellier: Éditeur du Montpellier Médical. A copy of Dr. Gachet's thesis, inscribed "to my friend and compatriot, A. Darimon," is held by the National Library of Medicine, Bethesda, Maryland (acquired September 12, 1892) and is available on microfilm reel 78–11, #2.

patient in an attempt to elicit a cure. Usually the doses were so small that nothing pharmaceutical was achieved. However, the homeopaths have been praised for their rejection of bleeding and purging, practices slow to die out among their allopathic contemporaries. Gachet may have retained a homeopathic interest but seems to have practiced allopathic medicine for the most part. This was but one aspect of his eclecticism.

Most commentators have labelled Gachet both eccentric and eclectic. The former adjective springs from his unusual jackets and caps, the rumor that he dyed his hair, the managerie of animals and birds at his house in Auvers-sur-Oise, and so on; the doctor was unusual, different, but never boring. "Eclectic" needs a little qualification. Today, as then, it implies a multiple approach, a choosing among various systems, doctrines or sources. In the modern context it invariably connotes a positive broad-minded approach but in the nineteenth century there was a slightly negative taint derived from the accompanying tendency to try any scientific island in a sea of uncertainty. When the skepticism was unorganized the eclectic resembled a dabbler or a dilettante.

Dr. Gachet in his early forties was already a free thinker, a socialist, and well connected with the world of arts and letters. He knew Courbet, Champfleury, and Victor Hugo and subsequently Paul Cézanne and nearly all the Impressionists. Paul Gachet met Cézanne's father in Aix-en-Provence, while completing studies in nearby Montpellier, and later interceded on behalf of the artist for more financial support. Gachet enjoyed the cafés frequented by struggling artists but was equally at home in the banquets of the *Société des Eclectiques,* founded by Aglaüs Bouvenne, wherein his wit and a penchant for vigorous discussion were noted. He printed the society's announcements, embellished with his own humor, on his own etching press.

Paul Gachet had an interest in the military, which started in Lille as a result of a relative who was an army doctor. He retained some contact through the Franco-Prussian War (1870), in which he served as a doctor attached to a mobile hospital. He had a brief flirtation with formal politics in 1878, when he ran in the municipal elections of Auvers-sur-Oise, on the conservative ticket along with the painter Charles Daubigny (1817–1878). Neither was successful.

He was fascinated by visual art (Figure 7.2) but equally intrigued by the creative mind. Dr. Gachet spent much time in 1866–68 with the etcher Charles Méryon, after the unfortunate artist had been committed at Charenton. A lasting interest in physiognomy, and in sketching patients, was later extended by offering formal instruction to students. On the other hand, a campaign for the establishment of a *Society for Mutual Autopsy,* to study the brains of creative artists, met with little enthusiasm. Auguste Renoir was one who resisted that invitation but nonetheless retained a

Figure 7.2. Dr. Paul Gachet [P. van Ryssel], 1887, etching, 11 × 9 cm (4.33 × 3.54″), portrait of Jean Baptiste Lamarck [1744–1829], French naturalist, collection of the author.

wholesome view of Gachet, who had nursed him through pneumonia in 1882.

Notwithstanding his thesis as exemplar of literary ability, Gachet favored the visual arts and wrote sparingly. However, to put the era in perspective it is worth mentioning a letter to the editor of *Le Figaro* (August 18, 1859) in which Gachet condemned their previous article claiming mental illness to be *contagious*. In a scientific paper of 1875, Paul Gachet spoke of current refinements in the humane treatment of the mentally ill and the possibilities of cure; he emphasized that the patient's freedom was indispensable for success.

Gachet's association with Vincent van Gogh was short but intimate. Their interaction has already been documented, via the letters, in earlier chapters. Dr. Gachet died January 9, 1909 of cardiac insufficiency in the care of Dr. Leon Simon, son of his old friend and mentor, Dr. Vincent Simon. He is buried in Père Lachaise Cemetery in Paris. He amassed a wonderful collection which was subsequently donated to the state by his children, Paul and Marguerite, in 1949, 1951, and 1954. It included 10 van Goghs, 8 Cezannes, 6 Guillaumins, 3 Pissarros, and single canvases by Renoir, Monet, Sisley, and Guys. The bulk of the collection is now on permanent exhibition in the Dr. Gachet room of the Musee d'Orsay. The value of the bequest on the open art market today would be in excess of $1,000,000,000.

Doctors in the sketchbook

Page twenty-four of sketchbook number two,[11] which was used by Vincent in Nuenen, Antwerp, and Paris, contains notations on books by Drs. Jozan and Capuron. These reminders were apparently written in Paris as judged by another entry, the address for Académie Roll, on the same page. Alfred Philippe Roll (1846–1919) was a French salon painter of some distinction who was noted for military themes and genre scenes. One of his pictures, *A Miner's Strike,* was lauded by Vincent (letters 238 and 280) and there were also passing references to Roll in letters 405, 406, and 615.

Antonie Émile Jozan (b. 1817) wrote a book, *Traité pratique des maladies des voies urinaires et des organes générateurs de l'homme,* (Practical treatise on the maladies of the urinary tract and the genital organs of man). The first edition was printed in Paris in 1850. From the second edition in 1851 and thereafter, the title was extended to "… men and women." Vincent sought the 14th edition, according to the note, and apparently added the library call number at a later date.

The second notation is to Capuron: *Traité des maladies des femmes, depuis la puberté jusqu'a l'âge critique inclusivement,* (Treatise on maladies of women from puberty to the critical age inclusively) first published in Paris in 1812. This book also went into several editions. Joseph Capuron (1767–1850) is mentioned by Ackerknecht[1] as a supporter of Pierre Charles Alexandre Louis (1787–1872) and his application of numerical methods to diagnoses. Capuron was an obstetrician at the Maternité, a maternity hospital transformed in 1795 from an old monastery in the Port Royal district of Paris, where he was interested also in the diseases of the newborn. He was regarded as one of the fashionable, great practitioners of Paris. Vincent's interest in these books remains a puzzle. It seems less likely that he sought information on a personal complaint but rather for some female acquaintance. Perhaps it was the mysterious "S" with whom Theo was embroiled in a relationship.

The doctors in Arles and St. Rémy

Félix Rey

Rey came to see the paintings with two of his friends, doctors, and they were uncommonly quick at understanding at least what complementaries are. I now intend to do a portrait of Monsieur Rey.
Vincent to Theo, letter 568, from Arles, January 1889.

Rey is a very nice fellow, a tremendous worker, always on the job. What men the modern doctors are!
Vincent to Theo, letter 585, from Arles, April 21, 1889.

Félix Rey was only 23, an intern at the Hôtel Dieu, in Arles when he first encountered Vincent van Gogh. Vincent took to him quickly and agreed with Theo that they ought to put Rey in touch with Dr. Rivet but "it would be a good thing to send M. Rey back here to [this] hospital with the doctor's degree he is trying to get ... Now Rey was born here [1867] ... once furnished with the full medical authority of Paris he would do real miracles here ... Rivet himself will perhaps be of the same opinion, at any rate in so far as an Arlesien is not a Parisian and vice versa" (letter 570). Rey did visit Paris but completed his degree at Montpellier, in 1890; his thesis was on the antisepsis of the urinary tract.

Dr. Aussoleil, a friend and colleague of Rey, was an intern at the nearby psychiatric hospital at Montdevergues and also took his M.D. at Montpellier in 1890. His thesis, *Quelques observations d'epilepsie larvée,* (Some observations of masked epilepsy), is supposed to anticipate temporal lobe epilepsy. This association is the proposed source of inspiration for Dr. Rey's "some sort of epilepsy" diagnosis for Vincent.[12]

In due course Vincent's promise to paint Dr. Rey's portrait (letter 568) was fulfilled. In a subsequent letter Vincent related how he had told Dr. Rey "to turn collector ... not try to paint himself" (letter 570). Both alternatives were rejected and Vincent's early enthusiasm for Rey's artistic education was apparently misplaced. The doctor was polite but disinterested in the painting, allowed his mother to use it to stop a hole in the hen house, and eventually sold the portrait in 1901 for a meagre sum. Nonetheless Dr. Rey maintained a semblance of correspondence with Vincent (through Theo) even after his patient left Arles, and the van Gogh brothers maintained a good opinion of him.

Rey's career was spent almost exclusively in Arles, apart from World War I service in Corsica. He then became the chief of medicine for Arles and the surrounding districts and was active[4] in dispensing tuberculosis vaccine (developed by Calmette and Guérin in 1921). He died there in September 1932.

Théophile Peyron

> *The director [Dr. Peyron] ... is a little gouty*
> *man – several years a widower, with very black*
> *spectacles. As the institution is rather dull, the*
> *man seems to get no great amusement out of his*
> *job, and besides he has enough to live on.*
> Vincent to Theo, letter 593, from St.
> Rémy, June 1889.

Dr. Peyron (1827–1895) made his first medical career with the navy and then settled in Marseille as an oculist. His appointment as director at the asylum of St. Rémy may well have been a semi-retirement position, as Vincent seemed to suggest in the above quote.

Patient and doctor had a roller-coaster relationship in the asylum. Early confidence in Peyron eroded and later comments by Vincent were always somewhat critical. "I should have stayed in St. Rémy another year if I had not attributed my last attack partly to the influence which the illness of others had on me ... (Dr. Peyron and I) had words over it, but we separated on good terms, and he has asked Theo for news of me" (letter 639, to Vincent's mother). He felt abandoned by Peyron: "my last attack, which was terrible, was in large measure due to influence of the other patients, and then the prison [asylum] was crushing me, and old Peyron didn't pay the slightest attention to it" (letter 648). There is some indication that Dr. Peyron did not prescribe bromides to the extent used in Arles, "it is odd that here [Auvers] the nightmares have ceased to such an extent; I always told M. Peyron that returning to the North would free me from it, but it is also odd that under his direction [in St. Rémy], though he is very capable and certainly wished me well, it was somewhat aggravated" (letter 640, June 10, 1890). Apparently Dr. Peyron did not send medical records north with Vincent, however, he did visit Theo once in Paris, at the end of September 1889 (letters T18, 608).

Ricord and Raspail

Everyone here is suffering from fever,
hallucinations or madness ... as for
considering myself completely sane, we must not
... People here ... [say that] there will always
be moments when you lose your head ... [but]
*the **Ricord** of this [subject] is probably **Raspail**.*
Vincent to Theo, letter 576, from Arles,
February 3, 1889.

Familiarity with both Ricord and Raspail demonstrates that the van Gogh brothers were aware of some of the important medical figures of the day. Ricord was a traditional, but hardly typical, man of medicine; Raspail was the contentious, antiestablishment advocate of better health for the masses. Although this is the only formal mention in the letters, Raspail takes on added significance because his book was incorporated into a painting by Vincent. That the artist put him in the same league as Ricord gives us a passing reference to the esteem with which Raspail's book of home remedies was held during the nineteenth century. In the quotation Vincent even accords Raspail expertise in mental health to the degree enjoyed by Ricord in venereal diseases.

Philippe Ricord

Ricord[13] was born at Baltimore on December 10, 1800 and at the age of twenty migrated to Paris where he was a student under the surgeons Baron Guillaume Dupuytren (1777–1835) and Jacques Lisfranc (1790–1847). He obtained his Doctorate from the Paris Faculty of Medicine in 1826. He first went to the provinces and started a successful general practice but returned to Paris in 1828. He became attached to the Pitié Hospital as surgeon and later was appointed principal medical officer at Hôpital du Midi which specialized in the treatment of male venereal diseases. Ricord soon became a worldwide authority in this area because of his lecture series, his practice, and his books.[14,15] Ricord offered the final proof in 1883 that gonorrhea and syphilis were separate entities. Ricord also classified syphilis into three stages: primary, secondary, and tertiary.

He had a clever footman who sped up the process by classifying the patients by sight as they entered Ricord's consultation offices. Ricord offered free services for those unable to pay. He was courteous, affable, full of wit. Oliver Wendell Holmes[16] described him as the founder of a remarkable school of venereology of France, the Voltaire of pelvic literature, a

214

sceptic on the morals of the human race, capable of submitting Diana to treatment with his mineral specifics or ordering a course of blue pills, mercurials, for the vestal virgins. Ricord died in Paris in 1889.

Raspail

> *How can I consider myself a doctor when all the world is going to become, without much pain, as learned as I?*
> Raspail, from the title page of his *Manuel Annuaire de la Santé.*

Thus wrote Raspail (Figure 7.3) on the title page of his annual book of health,[17] to the chagrin of most other physicians of the day. Francois-Vincent Raspail was born in 1794, at Carpentras (Vaucluse) in the south of France, actually only about 35 miles (56 kilometers) northeast from Arles. He was educated for the priesthood at the Seminary St. Charles at Avignon but with escalating skepticism he revolted against theology and left to become a schoolteacher and librarian back in Carpentras. He supported Napoleon Bonaparte and the eventual fall of the Emperor was a heavy blow. After self-analysis he started "doubting everything and subjecting everything to the compass of reason and demonstration" and mentioned in his autobiography "an insatiable desire to fathom the depths of science."[18] In 1816 he left for Paris.

Raspail became one of the most unique, popular, and influential figures in nineteenth century France. He had already made significant contributions to chemistry, biology, and particularly histochemistry when he turned his attention to the health and welfare of the general population. His system stressed hygiene, focused on common medical problems, and depended on a few chemicals such as camphor (unfortunately touted as a panacea) and a handful of crude plant extracts. He published his *Manuel Annuaire de la Santé* in pocket-book format starting in 1845, and sales averaged over a hundred copies a day in the first five years.[18] It was repeated every year but two until his death, in 1878, and then it was continued by his descendants till 1935. Raspail's contentiousness extended into politics, which landed him in jail for more than two years and exile for nine, but his publications, speeches, and ministrations elevated his recognition to a household name.

Vincent was preoccupied with his own health and even wrote his brother, as early as July 1880, about the wisdom of self-study in medicine (letter 133), and expressed some organized skepticism about prevailing norms of health care (letter 148). Attraction to the low-cost, home-remedies of Raspail (letter 576) by someone of Vincent's demeanor is easily rationalized.

Figure 7.3. Portrait of F.V. Raspail, engraving, after 1863 edition of *Manuel Annuaire de la Santé.*

Indeed, he lauded Raspail and incorporated an image of the *Manuel* into *Still Life: Drawing Board with Onions, etc.* F 604, which was painted between January 8 and 17, 1889. During the same period (shortly after leaving the Hospital at Arles) Vincent wrote about his enthusiasm for camphor in overcoming insomnia (letter 570); this was surely inspired by the book, although Vincent's reckless "... very, very strong dose ... in my pillow and mattress" can be contrasted with Raspail's directive of chewing about 50 mg at night. (Raspail recommended camphor also as a remedy for nymphomania, priapism, and satyriasis, and it was actually for "the arrest of the precocious habits of childhood" that he spoke of sprinkling camphor powder in the bedding.) Van Gogh had a copy of the *Manuel,* considered it worthy of a legible title in his painting, and joined thousands of compatriots in consulting this book of home remedies.

Concluding remarks

> *I like to talk with a doctor now and then ... if*
> *one occasionally hears a sound and true word*
> *about health, ... (one has a) clearer idea ...*
> *(and is) not shaken like a reed by every wind*
> *and does not believe all the nonsense so often*
> *heard about health and ill health.*
> Vincent to Theo, letter 148, from Etten,
> mid-1881.

Vincent was assisted into this world in 1853 by Dr. Cornelis van Ginneken of Zundert,[4] and attended on his deathbed in Auvers-sur-Oise by Drs. Jean Mazery and Paul Gachet. In the intervening 37 years he met many others including Dr. van Gent, in Etten, in 1881, whom Vincent described as a "clever and practical man" (letter 148), and Dr. Van der Loo, in Eindhoven, in 1885, from whom he extracted an expectation of living to sixty or seventy (letter 442). The doctors in The Hague have not been identified although one at least was the recipient of a van Gogh painting. Drs. Rivet, Rey, and (to a major and enthusiastic extent) Gachet were also owners of van Gogh paintings. We should also mention Christian Mourier-Petersen (1858– 1945), he of the quote, "it's the doctors who kill people" (letter 490). He was an unsuccessful medical student who befriended Vincent in Arles and then Theo in Paris. He returned to his native Denmark, where his success was modest as an artist, and somewhat greater as an associate of Vincent van Gogh.

Drs. Gruby and Ricord were outstanding scientists by any standard. Raspail is a wonderfully interesting figure, whose fame derives from his book of health, while his scientific reputation rests primarily on pioneer work in histochemistry. Dr. Gachet's interaction with other artists such as Cezanne, Renoir, and Pissarro would have assured him a place in history. However, there can be little doubt that Gachet's brief relationship with Vincent van Gogh engendered most of the biographical interest in the doctor, and essentially determined the careers of his son and daughter.

There is neither cause nor justification for casting stones at Vincent's doctors. Notwithstanding some attempts in the older literature to blame the physicians for Vincent's problems, and an absurd revival of criticisms directed at Dr. Gachet for so-called exploitation of van Gogh, the overall assessment is rather that they did as well as expected with an unfamiliar disease and a difficult patient. Cases of acute intermittent porphyria escape detection even today. The occasional survey of mental institutions reveals a significantly higher incidence of acute intermittent porphyria than that discovered in the general population. Likewise, patients eventually documented with this problem often show abdominal scars resulting from the unproductive exploration of hasty surgeons.

References

1. Ackerknecht EH. 1967. *Medicine at the Paris Hospital 1794–1848*. Baltimore: The Johns Hopkins Press.
2. Pelicier Y. 1975. France. chapter 4, in: *World History of Psychiatry*. JG Howells (ed.) New York: Brunner and Mazel.
3. Jones RM. 1967. An American medical student in Paris, 1831–1833. *Harvard Library Bulletin* 15: 59–81.
4. Gachet P[L]. 1957. Les médecins de Théodore et de Vincent van Gogh. *Æsculape* 40: 2–37.
5. Hulsker J. 1990. *Vincent and Theo van Gogh: A Dual Biography*. p 454. Ann Arbor: Fuller Publications.
6. Kruta V. 1972. David Gruby. vol 5, pp 565–566, in: *Dictionary of Scientific Biography*. CC Gillispie (ed.). New York: Charles Scribner's Sons.
7. Zakon SJ & Benedek T. 1944. David Gruby and the centenary of medical mycology. *Bulletin of the History of Medicine* 16: 155–168.
8. Anonymous. 1898. Obituary: "An eccentric physician." *The British Medical Journal* vol 2 for 1898: 1705–1706.
9. Doiteau V. 1923. La curieuse figure du Dr. Gachet. *Æsculape* 1923: 169–173, 211–216, 278–283.
10. Fabbri R Jr. 1966. Dr. Paul-Ferdinand Gachet: Vincent van Gogh's last physician. *Transactions and Studies of the College of Physicians of Philadelphia* 33: 202–208.
11. van der Wolk J. 1987. *The Seven Sketchbooks of Vincent van Gogh: A Facsimile Edition*. p 110. New York: Harry N. Abrams Inc.
12. Gastaut H. 1956. La maladie de Vincent van Gogh envisagée à la lumière des conceptions nouvelles sur l'épilepsie psychmotrice. *Annales Medico-Psychologiques* vol 1 for 1956: 196–238.
13. Waugh MA. 1990. History of clinical developments in sexually transmitted diseases. chapter 1 in, *Sexually Transmitted Diseases*. 2nd ed. KK Holmes, P-A Mårdh, PF Sparling and PJ Wiesner (eds). New York: McGraw-Hill Information Services Company.
14. Ricord P. 1842. *Traité Pratique des Maladies Vénériennes ou Recherches Critiques et Expérimentales sur L'Inoculation Appliquée a L'Etude de ces Maladies, Suives d'un résumé thérapeutique et d'un formulaire spécial*. (translated by HP Drummon). London: Longman, Brown, Green & Longmans. (French edition, 1838, Paris: Librairie des Sciences Médicales de Just Rouvier et E Le Bouvier.)
15. Ricord, P. 1851. *Illustrations of Syphilitic Disease*. (translated by TF Betton) Philadelphia: A Hart.
16. Holmes OW. 1888. *Medical Essays 1842–1882*. p 437. Boston & New York: Houghton Mifflin and Company.
17. Raspail FV. 1886. *Manuel Annuaire de la Santé pour 1886*. Paris: Chez l'Editeur. (This is the edition which is available in Kansas City.)
18. Weiner DB. 1968. *Raspail, Scientist and Reformer*. New York & London: Columbia University Press.

Chapter 8
The Yellow Palette

Figure 8.1. Vincent van Gogh: *The Night Café*, 1888, Arles, oil on canvas, 70 × 89 cm (27.6 × 35″), Yale University Art Gallery, New Haven, Connecticut, Bequest of Stephen Carlton Clark

*The more ugly, old, vicious, ill, and poor I get
the more I want to take revenge by producing a
brilliant color, well arranged, resplendent.*
Vincent to Wil., letter W7, from Arles,
September 8, 1888.

Instant recognition of the image is soon followed by an appreciation of the palette. These are two of the many aspects which contribute to the popularity of Vincent van Gogh's paintings in museums, and explain their prevalence as decorative reproductions. From the Paris period starting in 1886 until his death in 1890, Vincent created hundreds of paintings with vibrant colors, and on several occasions made a formal commitment to color. Thus, in the second half of 1886, he wrote to an English artist, Horace Livens: "In Antwerp I did not even know what the Impressionists were, now I have seen them and, though not being one of the club yet, I have much admired certain impressionists' pictures ... [as for myself] I have made a series of color studies in painting ... seeking oppositions of blue with orange, red and green, yellow and violet, seeking ... to harmonize brutal extremes ... to render intense color and not a grey harmony ... I have faith in color, even with regard to the price the public will pay for it in the long run" (letter 459a).

In Arles, March 1888, Vincent declared that he believed "in the absolute necessity of a new art of color" (letter 469). In his following letter to Theo we see the first mention of the use of color in bold design "like stained glass windows." It was a visual approach based on the philosophy of Guy de Maupassant for creative writing, to wit, "the artist's liberty to exaggerate, to create in his novel a world more beautiful, more simple, more consoling than ours" (letter 470). A little later, we see another breakthrough for Vincent, "Many subjects here are exactly like [those in] Holland in character, the difference is in the color. There is that sulfur-yellow everywhere [that] the sun lights" (letter 488). In mid-1888, he was convinced of "the importance of staying in the Midi, and of positively piling it on, exaggerating the color" (letter 500). In this chapter we will explore the physical and chemical background to Vincent's remark, "How lovely yellow is!" (letter 522), and examine the dominance of the yellow and the red regions of the spectrum in many of his paintings. Our primary assessment[1] of some of the subject matter of this chapter appeared in 1991.

Several working hypotheses posit medical problems having some influence on Vincent's art; analyses of neurological, nutritional, environmental, and chemical factors have all received some discussion. These elements influenced his life, and some of them probably shortened it, but did they affect his view of the world, his style, his palette?

On the one hand, Arnheim[2] was chagrined by colleagues who attempted to find medical problems in van Gogh and other artists. He felt that such explanations were "often sought [only] to support the conviction that if those artists had not been physically or mentally impaired, they would have produced [more] realistic pictures." He enjoined us to explore only symbolism by allowing that "it makes sense to inquire what the color yellow meant to Vincent van Gogh in 1889 or what blue meant to Picasso in 1903." On the other hand, we should not lose sight of the crux of the matter; that is, does the artist's personal view have much to do with the final product? Most would say that it does and it seems to me equally reasonable to wonder if transient or progressive abnormalities in vision are also influential?

Perception versus rendition

If a painter suffers from diplopia (double vision), and sees a distant telegraph pole in the landscape as two vertical lines, then he should still be content to make a single stroke of the brush on the canvas because this line will also appear to him as a doublet. So goes the cardinal objection to any theory invoking an effect of eye disease on an artistic image. In the words of Trevor-Roper,[3] "the artist paints what he sees, and the subject will correspond to the rendering, however much they are both altered by the misshapen eye into a distorted percept within the artist's occipital lobe." Medawar[4] went so far as to say that anyone who would suggest otherwise was decidedly dull, and he joined the throng who roundly criticized a suggestion of astigmatism in El Greco.*

However, if our hypothetical painter is primarily myopic (short-sighted) and perceives a blurred double image of that same distant pole, then he would realistically use two strokes on his canvas, which is but arm's length away, and is essentially in focus. Trevor-Roper[3] made this distinction and went on to say that "when a naturalistic artist suffers from certain eye diseases such as myopia, presbyopia (diminution of the power of accommodation) or color-blindness his renditions must surely be affected." Mills[5] made a good case for myopia in several of the successful Impressionists. That author, both myopic and astigmatic himself, also offered a charming and instructive description of his uncorrected view of a concert orchestra.

Another feature to both geometric and color distortion depends entirely

---

* Modern evidence[3] from X-ray photographs has put that particular case to rest by showing that the elongations in some parts of El Greco's paintings were secondarily imposed on the original sketches.

upon the evaluation of the experience. Thus an artist may view a scene under abnormal conditions which lead to aberrancies, be inspired by the new view, and then decide to reproduce the exotic image *under normal conditions*. The possible causes of such transient aberrancies are manifold but include intentionally removing one's spectacles, looking at the world through the bottom of a whisky glass, and ingesting chemicals which induce reversible chromatopsias (distortions in color perception).

Color perception

Humans have a color response to wave lengths of light between 400 and 700 nm.[*] Color vision is an integrated process that starts with the generation of signals in retinal receptors, involves comparisons and evaluations of the information that is transmitted to the brain, and terminates with the declaration of particular hues in order to describe a scene. Physical, chemical, physiological, and psychological aspects are all involved.

When a beam of light from an incandescent lamp passes through a prism it emerges as a spectrum, the so-called colors of the rainbow. The specific band (region) of the spectrum which we perceive as green is made up of light with a small range of wave lengths centered around 510 nm. The following table shows the relationship between "psychologic" color and the wave length of incident light.[6]

Violet	Blue	Cyan	Green	Yellow-green	Yellow	Orange	Red
450 nm	470 nm	495 nm	510 nm	560 nm	575 nm	600 nm	660 nm

Unique aspects to yellow

Color discrimination, the ability to discern one spot of light superimposed on a larger one of slightly different wave length, is most sensitive in the yellow region of the spectrum.[7] These experiments were performed with a black background. Other hues are discriminated with maximum fidelity against a yellow background.[8] Land[9] postulates that the key to color vision is the comparison of information of longer versus shorter wavelengths and he calls yellow "the fulcrum in the ordinary sunlit world," all of which may help explain why van Gogh of the "Potato Eaters" period, Rembrandt, and

---

[*]   A nanometer (nm) is 0.000000001 meter.

other masters before them arrived at a keystone of yellow. A Kansas City artist told me that a disappointing color composition can sometimes be "pulled together" by a glaze of yellow over the whole canvas. This review should be sufficient to indicate that there are special aspects to yellow and it is not just symbolism.

Ageing of the human lens

The perception of the full range of colors is subject to a multitude of influences. One natural example concerns ageing. The fraction of incident light energy which reaches the retina decreases with age; this is primarily

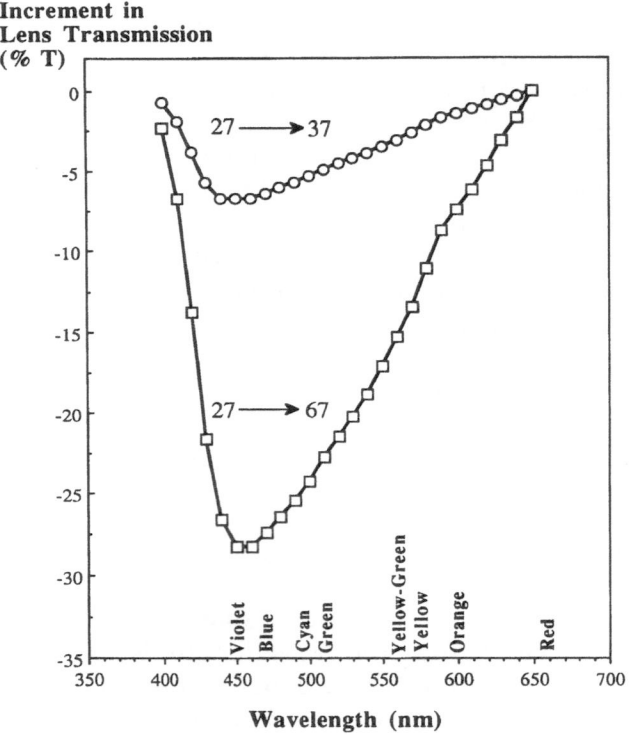

Figure 8.2. Normal ageing of the human lens. The decrease in percent transmission is plotted against the wave length of incident light. The upper curve compares a 37 year old lens with a 27 year old lens; this simulates the small degree of "yellowing" of Vincent's lenses during his artistic career. The lower curve compares a 67 year old lens with a 27 year old lens, showing a more significant decrease in the percent transmission (particularly of violet-blue light) due to ageing. Data were calculated using the equations of Pokorny et al.[11]

due to a gradual loss in transmittance by the lens.[10] The attenuation is more pronounced for the shorter wave lengths and is maximal between 450 nm and 470 nm, which are centered about the color responses of violet and blue respectively. This may be the explanation for less blue and more grey and green in the skies of older landscape painters.[3]

Pokorny et al.[11] summarized the available data on healthy eyes of different ages, and with their equations we calculated that the loss in lens transmission between ages 27 and 67 is a significant 28% at 460 nm, as shown in Figure 8.2. But for Vincent, over his artistic career (27–37 years), the predicted change for healthy eyes is less than 7%, and this degree of "yellowing" of the lens is insignificant compared with the extent that attends the unnatural condition of xanthopsia, a pathology of the color sense which elicits yellow vision. The effect is usually reversible, but patients or volunteers have described their experiences as if they were looking through yellow filters.

Experiments with yellow filters

In order to simulate xanthopsia we used yellow filters cut from No.8 Kodak® gelatin sheets and adapted to a pair of welder's goggles. This filter has greater than 90% transmittance between 700 nm and 600 nm, 84% at 520 nm, 0.25% at 460 nm, and is virtually opaque to shorter wave lengths, i.e. it cuts out the wave lengths of light that elicit a blue response. We used cut-outs of colored papers to construct three small pictures depicting simple scenes with easily recognizable features. Medical students with normal color-vision were individually fitted with the goggles and then asked to view the pictures in a room with indirect sunlight. They were asked to name the colors of items in a standard order. The table lists the anticipated color distortions.

Without Filter	With Yellow Filter
light blue	grey
medium blue	green
dark blue	black
purple	brown
white	pale yellow
green	green
yellow	yellow

Distortions in perceived colors due to the yellow filter were generally as we had anticipated. We encountered a small number of interesting exceptions which we believe can be explained by a phenomenon called color constancy.[12] Some subjects apparently received a cue from the white notepad of the interviewer and "adapted" to the yellow filter. For example, 18 subjects reported the purple window frame in picture #1 as brown (with the filters before their eyes) compared with 2 subjects who still described it as purple. For the same purple and medium-blue papers that were incorporated into all three pictures, the degrees of color constancy (in contradistinction to reporting brown and green respectively) within the group of viewers increased slightly from picture #1 through pictures #2 and #3, which we attribute primarily to increased cues with time and appropriate adaptation.

We also asked six professional artists to don the goggles and attempt to match the colors of one of the pictures by mixing primary pigments. (They had not seen the picture or the blank canvas without the filters.) In each case we asked them to name the colors before commencing and then to reproduce the scene. All artists complained of difficulty in achieving the "right" color, but otherwise the results were variable. One artist described the predicted color shifts mentioned previously and then proceeded to mix the colors in a manner that seemed greatly influenced by those declarations of distorted colors. Another artist enunciated the same color shifts but mixed pigments that were very reasonable matches to the original picture viewed without filters. But, before removing the goggles, he tried to match a particularly difficult "blue" (which he had declared a "green") with a dab from a commercially-mixed tube that he was familiar with by label. We (and later the artist) were all surprised that the disparity was actually *increased*.

The camera lacks the cerebral activity behind color constancy and photographs of our experimental pictures show all of the anticipated color distortions. The same is true for reproductions of art work. Recall that Vincent van Gogh's *The Night Café* (F 463) and Paul Gauguin's *In an Arles Café*, depict the same room, but the latter has a much cooler palette. A color transparency of Gauguin's picture photographed through the yellow filter was received by a lecture audience in Kansas City as "more like a van Gogh." Our little operation is reminiscent of an observation by Goethe,[13] "if we look at a landscape through a yellow glass, particularly on a grey winter's day [then] the eye is gladdened, the heart expanded and cheered, a glow seems at once to breathe towards us."

The students' interpretations of blue tones were distorted, and the artists had difficulty mixing and matching colors under the influence of the yellow filters. These experiments were designed primarily for our own amusement, and I hasten to add that there is a large body of literature which addresses the complexities of color perception in much more controlled and sophisticated ways. But even unsophisticated trials such as ours are sufficient to indicate the possibilities for confusion in color perception under a simulated xanthopsia. Before examining some of the chemical and physical insults that can induce xanthopsia we should consider the nature and extent of Vincent's exaggerated use of yellow pigments.

van Gogh's "high yellow" paintings

From 1886 to 1890 van Gogh produced at least 638 paintings.[14] Color reproductions of all items from this period are not readily available, but an evaluation[1] of 276 illustrations from seven reference books,[15–21] provided a reasonable survey of the extent of yellow dominance. We avoided duplication and excluded self-portraits which were dealt with separately. The paintings in this survey were produced in Paris (20%), Arles (39%), St. Remy (24%) and Auvers-sur-Oise (17%). We define *high yellow* pictures as being rich in yellows, but virtually lacking blues, violets, and white, for the reasons discussed in the previous sections. We found 10% in this category. Examples are given in the table.

Table 7. Examples of the three types of palette

	Painting	de la Faille[14] number	Venue	Year
I.	"High Yellow" [10%]			
	Sunflowers	F 376	Paris	1887
	The Night Café	F 463	Arles	1888
	Roses	F 681	St. Rémy	1890
	Trees	F 817	Auvers	1890
II.	"Blue Swath" [30%]			
	The Bathing Boat	F 311	Paris	1887
	Wheat Field	F 411	Arles	1888
	The Reaper	F 618	St. Rémy	1889
	Sheaves of Wheat	F 771	Auvers	1890
III.	"Full Spectrum" [60%]			
	Restaurant Interior	F 342	Paris	1887
	Café Terrace by Night	F 467	Arles	1888
	Starry Night	F 612	St. Rémy	1889
	Crows over the Wheat Field	F 779	Auvers	1890

There were other canvases within this sample which are rich in yellows, greens, and reds but have a swath of blue or violet, these were defined as *blue swath*, about 30% fell into this category. The warmer hues pervade many of the remainder, but they have a more balanced palette that includes the blue end of the spectrum, these were counted as *full spectrum*. Within the full sample only 5% were judged to have a somber palette. Bonafoux[22] assembled 37 self-portraits by van Gogh. We deem 8 of these portraits to be *high yellow* paintings and only one, *Self-portrait dedicated to friend Laval*, F 501, 1888, was painted in the South.

The survey has obvious defects, not least of which are the variation in the quality of color reproductions from one book to the next. Notwithstanding several caveats, including the artist's "touching-up" paintings at a later date, the following conclusions are merited. The great majority (95%) of the paintings in this survey were made with a bright palette. High yellow paintings were significant in *number*, their execution was *episodic*; and they were produced at *multiple locations*. There is no indication of a continuous trend toward a more yellow palette. The findings could be accommodated by occasional, reversible xanthopsia.

We know that this condition can distort color perception and confuse selection of pigments. Also, the impressions under xanthopsia may be rendered later, from memory, after return to normal color vision. In either case the choices of pigments may seem unrealistic, exaggerated, or exotic to observers and even to the recovered artist. Walsh and Hoyt[23] mention at least thirteen chemicals that can cause xanthopsia. We can exclude some immediately because they are twentieth century drugs. Others include digitalis and santonin, which were used in France during the period of interest, and are worth reviewing. We will also discuss absinthe drinking and sunstroke in this context.

The approach

In the sections that follow, most of the chemical and physical insults that can lead to xanthopsia are introduced one at a time. Each is examined for relevancy to Vincent van Gogh, and the available data are evaluated for either support or denial of each working hypothesis. Colleagues who listened to my explorations on this subject likened the process to erecting straw men (each working hypothesis) and then knocking them over (with the data). The individual results are not always exciting; that is the nature of research. I realize that an invitation to worry along with the author is not common in a world preoccupied with end results. Nonetheless, I trust that the exercise is instructive and that the items have intrinsic appeal.

Digitalis intoxication

Withering's book (1785) on the medical uses of *Digitalis purpurea** warned that, "the foxglove when given in very large and quickly repeated doses occasions sickness, vomiting, purging, giddiness, *confused vision, objects appearing green or yellow*" (my italics).[24] The following is a précis of Withering's case # 106:

> A 61 year old man did not respond to previous treatment with squills [a preparation from the bulb of *Urginea maritima* containing cardioactive glycosides] for hydrothorax [a collection of watery fluid in the pleural cavity] and swelled legs. I [Withering] was apprehensive upon my second visit that he had not 24 hours to live [so] ... I gave him the *infusum digitalis* [a decoction of *Digitalis purpurea*] stronger than usual ... Finding himself relieved by this he continued to take it, contrary to the directions given, after the diuretic effects had disappeared. The sickness that followed was truly alarming: it continued at intervals for many days, his pulse sank to forty a minute, every object appeared green to his eyes, and between the exertions of retching he lay in a state approaching to syncope [temporary suspension of consciousness] ... At length, however, he did begin to emerge out of the extreme danger into which his folly had plunged him; and by generous living and tonics, in about two months he came to enjoy a perfect state of health.

Johannes Purkinje (1789–1869) described his own symptoms after ingestion of a concentrated decoction of digitalis.[25] He was especially vivid in the illustration of a scintillating scotoma (a luminous, sparkling zigzag outline around an area of lost vision). His diagrams are reminiscent of the drawings by Hubert Airy for scotomata associated with migraine (1870), which were reproduced by Duke-Elder.[26] Purkinje noted that during the entire time of his intoxication with digitalis "the brain was not affected in the least (as, for instance, after opium, camphor or datura)."[25] Good old Purkinje was a devil for punishment!

In this century, Sprague et al.[27] described seven cases involving chromatopsias from digitalis intoxication; four had yellow vision, one red-yellow, and two green. Lely and Van Enter[28] reported a large-scale digitoxin intoxication in 1969 due to an error in tablet manufacture, and they observed serious eye conditions in 170 (95%) of the patients. As well as

---

* The foxglove plant, *Digitalis purpurea*, is the source of the drug digitalis, a mixture of glycosides of which digitoxin is the most active principle.

episodes of amblyopia and diplopia, and experiences of light flashes and scintillating scotomata, patients and volunteers complained of aberrancies in color vision wherein all objects appeared primarily green, yellow, or white. It is characteristic of digitalis intoxication that the latter symptoms disappear in a few days after withdrawal from the drug.[28,29] There is no unanimity on mechanism, but Gibson et al.[30] proposed that the yellow vision of digitoxin toxicity is related to a selective, reversible effect on receptors rather than the optic nerve.[31]

Robertson et al.[32] emphasized that patients are generally reluctant to admit to distorted color vision, thus the statistics are possibly underestimated. Nonetheless, a large body of literature covering 200 years supports the fact that digitalis intoxication can induce xanthopsia. Lee[33] was the first to propose that van Gogh may have been treated with this medicine for a short period, noticed the golden and coronal symptoms, and, when these side effects disappeared, purposefully continued to paint with a yellow dominance. Rentchnick[34] and Lanthony[35] found some support for this working hypothesis and joined Lee[33] in reminding us that digitalis was given for epilepsy, mental illness, and other medical problems in the nineteenth century, albeit inappropriately by modern standards. There is no record of Vincent taking the drug, but these authors were intrigued by the sprigs of foxglove that Vincent incorporated into two portraits of Dr. Gachet, Auvers-sur-Oise, F 753 & 754, 1890. We[1] are more inclined to interpret the plant as a symbol for the physician rather than as a drug for the artist. Ravin[36] noted that because Gachet was a homeopathic physician he was less likely to have administered a toxic dose of digitalis. In any event, Dr. Gachet's contact (two months before Vincent's death) was late; if Drs. Peyron (St. Remy), Rey (Arles), Rivet (Paris), or Gruby (Paris) had been similarly depicted, then a more meaningful (but still circumstantial) case could have been made.[*]

There is also the possibility of self-administration of drugs by van Gogh. Again, we find nothing about digitalis in his correspondence, although he mentioned taking potassium bromide at Arles (letter 574). An indirect case can be made for santonin because it was recommended by Raspail,[37] whose system of medicine[38] was known to the artist (letter 576 and references 37, 39).

---

[*] Some popular-media reporters, supposedly in their zeal to offer a long list, have included digitalis intoxication among Vincent's major medical problems. The facts about Vincent's illnesses (chapter 3) offer no support for this idle suggestion. The possibility of digitalis intoxication playing any role in van Gogh's crises is so remote that it did not merit raising under other hypotheses (chapter 6).

van Gogh and Raspail

Vincent was preoccupied with his own health and even wrote to his brother, as early as July 1880, about the wisdom of self-study in medicine (letter 133), and expressed some organized skepticism about prevailing norms of health care (letter 148). Attraction to the low-cost, home remedies of Raspail (letter 576) by someone of Vincent's demeanor is easily rationalized. Indeed, he lauded Raspail and incorporated an image of Raspail's book into *Still Life: Drawing Board with Onions etc.*, (F 604) which was painted between January 8 and 17, 1889. During the same period (shortly after leaving the Hospital at Arles) Vincent wrote about his enthusiasm for camphor in overcoming insomnia (letter 570); this was surely inspired by the book, although Vincent's reckless "very, very strong dose ... in my pillow and mattress" can be compared with Raspail's directive to chew about 50 mg at night.[*]

Van Gogh had a copy of Raspail's *Manual of Health*, considered it worthy of a legible title in his painting, and joined thousands of compatriots in consulting this book of home remedies. We do not know which edition Vincent had, but I consulted the '63, '72, and '86 versions wherein appeared the same entry (pp. 133–134 of the 1886 edition[38]) on *semen-contra* (cf. *santonica* in U.S. Dispensatory[40]), i.e. the unexpanded floral heads of Levant wormseed which contained santonin as the active principle.

Santonin

This sesquiterpene lactone (Figure 8.3) is found in several *Artemisia* species. The commercial source is primarily *A. maritima*.[41] Santonin, as a crude decoction or powder and later as pure crystals, was used effectively as an anthelmintic for several centuries until displaced by drugs with less side effects. Xanthopsia due to santonin overdosage was probably encountered almost as soon as its efficacy against ascariasis was established and was documented[42] as early as 1806. It may also be important (cf. the *blue swath* palette) to note that while bright objects are yellow, dark surfaces sometimes have a violet appearance under santonin xanthopsia.[43,44] The visual symptoms may occur in the absence of all other toxic manifestations, which in extreme cases include convulsions.[44]

---

[*]   Raspail also recommended camphor as a remedy for nymphomania, priapism, and satyriasis, and it was actually for "the arrest of the precocious habits of childhood" that he spoke of sprinkling camphor powder in the bedding.

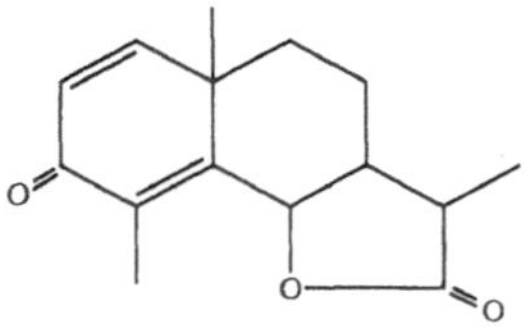

santonin

Figure 8.3. Santonin

A single dose of 200 mg santonin (about the upper limit of the therapeutic range) was sufficient to cause yellow vision in some individuals after two hours; with 500 mg or more the effect was noted in a half hour.[43] Duke-Elder[29] reported that doses as small as 100 mg santonin commonly caused yellow vision. Raspail suggested up to 2 g daily of *semen-contra* as required, this amounted to about 80 mg santonin [my calculation], although the concentration varied with the dried herb. More importantly, the administration of powdered products was often imprecise in Vincent's day. Raspail's direction, "a pinch with three fingers equals one gram" is a bit rough.* The large variations in actual amount taken, the small difference between therapeutic and toxic doses, and individual differences in sensitivity to santonin are sufficient to account for the incidence of intoxication.[43]

The drug was also taken in the nineteenth century as a preventative medicine or in response to vague and indefinite diagnoses; more recently documented examples involved individuals suffering dyspepsia and other gastrointestinal complaints who self-administered,[45] or were given by relatives,[46] santonin-containing preparations. Van Gogh frequently complained of gastrointestinal problems and may have suspected worms; his penchant for excess, as exemplified with camphor, may have led to overdosage on another terpene, santonin. Arnold[47] has suggested that van Gogh's affinity for absinthe developed into a pica for terpenes, the documented examples being thujone, camphor, and pinene. However, there is no direct evidence for van Gogh taking santonin.

Absinthe

Indulgence in alcohol added to van Gogh's illness and the injurious effects of absinthe in particular have been emphasized.[47,48] This liqueur was very popular in nineteenth century France; the per capita consumption was

---

* I found the standard deviation for 10 pinches of a dried plant powder was equivalent to 22% of the mean; the actual weight varied fivefold for different materials, and by as much as 50% depending on how finely they were ground.

particularly high in Paris and Arles in van Gogh's time,[49] and there are several indications that he developed an affinity for it (see chapter 4). He also painted *Absinthe Glass and Carafe*, F 339, Paris, 1887, and included a green (absinthe?) bottle into *Still Life: Drawing Board with Onions*, F 604, Arles, 1889. Vincent said that he painted *The Night Café*, F 463, on the spot, staying up three nights in a row and sleeping during the day (letter 533). It is tempting to speculate that he had a glass or two during the execution of this painting; he certainly had access, and the landlord was apparently pleased with the whole event. Vincent surmised that Tersteeg (the art gallery manager in The Hague) would surely judge Vincent as having "delirium tremens" while painting this picture (letter 534). Apart from the possibility of this special case, we do not imply that van Gogh painted while intoxicated. We feel that his creations occurred while lucid, but that novel experiences of relative sizes, shapes, and colors perceived under the influence of absinthe may have been recalled later and incorporated into new and daring compositions, perspectives, and palettes.

Absinthe contained high concentrations of both alcohol and essential oils (from herbs); the latter contributed to flavor, fragrance, and toxicity.[50] The most deleterious constituent was thujone, which can cause visual hallucinations.[51,52] In the 1920's and 1930's, thujone- and camphor-induced convulsions were studied as models for epilepsy (chapter 4, ref. 50). Subsequently camphor, Metrazole (pentylenetetrazole) and Indoklon (hexafluorodiethyl ether) were used successfully as pharmacologic agents to induce convulsive seizures in certain cases of schizophrenia, i.e. chemical forerunners of electroconvulsive therapy.[53] Therapeutic doses of Metrazole commonly evoked xanthopsia[54] or varicolored visual hallucinations[55] in the preconvulsant period (aura). Muskens[56] described a girl who suffered fits that were attributed to head trauma and notable because they were preceded, for several days on end, by visual hallucinations wherein she saw the landscape as extraordinarily gay in color. The epilepsy was successfully treated with bromides, but at age 20 she started occasional drinking, which precipitated more fits.

Notwithstanding chemical differences between Metrazole and the terpenes (camphor, thujone and santonin), all can act as convulsants. That Metrazole and santonin cause xanthopsia is also intriguing. Amblyopia and optic neuritis were reported under absinthism,[44] but chromatopsia was not encountered in our literature search.[1] Hilbert (1913) wrote about a young, neurasthenic, male patient who suffered a three hour bout of xanthopsia after an alcoholic excess, but the liquor was not specified.[57]

The herbs incorporated into absinthe varied with the manufacturer but always included wormwood *(Artemisia absinthium)* and Roman wormwood *(Artemisia pontica)*,[50] and we wondered whether they contributed santonin (cf. *Artemisia maritima*). The manufacturing procedure, as well as the low concentration of santonin reported for *A. absinthium*, excused that species. Our own analyses of an alcoholic extract of *A. pontica* (i.e., simulation of the secondary process in absinthe manufacture) indicated that the amount of santonin was insignificant.[58] Nonetheless, van Gogh's overindulgence in absinthe, together with an increased susceptibility to its toxic effects due to inadequate diet (chapter 3) and his underlying disease (chapter 5) may have combined to influence his perception on specific occasions, especially "after spending the whole day in the blazing sun" (letter item A16).

"Too much sun for a Nordic head"

We quote from the mid-1930's doctoral dissertation of Beer[59] who attributed this diagnosis to Dr. Paul Gachet, without documentation. A decade earlier, Grey[60] had assembled 33 reproductions of van Gogh's paintings, and remarked in the preface that the artist had, "suffered from a chronic kind of sunstroke." Again, no supporting evidence was provided and the idea was mentioned, but not seriously regarded, by subsequent reviewers.[15,33,61,62]

Vincent remarked upon being "dazed with the sun" (letter 512), that "beats down on one's head ... [and] makes one crazy" (letter B15). He noted that sunburn was common around Arles (letter 498a), and commented upon his own sometimes reckless exposure (letter B7). Our survey indicated that *high yellow* paintings were not restricted to the South and included some indoor scenes, still lifes, and self-portraits so that sunstroke (heatstroke) or heat exhaustion could not be the sole contributing factor.

Sunstroke results from uncontrolled elevation of body temperature after the reflex sweating system is over-taxed by environmental heat.[63] The syndrome has been recognized since biblical times but despite modern, aggressive attempts at management, the morbidity and mortality rates can still be impressive. According to an 1886 handbook,[64] vision is sometimes affected by heatstroke; 60 cases of chromatopsia were observed in New York City about this time. It is curious that more recent reviews of heatstroke do not mention chromatopsia; perhaps the syndrome is included under delirium or goes unreported because of the stigma attached to admitting to hallucinations.[32]

Duke-Elder[29] and Carroll[65] were of the opinion that the visual disturbances due to digitalis intoxication, and some other chemically invoked chromatopsias, are actually hallucinations, i.e. due to central function impairment caused by the drug. This working hypothesis has the charm of being able to embrace such disparate causes of xanthopsia as digitalis, santonin, and sunstroke but otherwise remains in conflict with hypotheses

that depend more on specific chemical or physical effects on photorecep-
tors.[31]

We encountered other chemical causes of xanthopsia,[23,44] but they were
dismissed because they were either isolated and exotic, e.g. picric acid,
chromic acid, carbon disulfide, oil of wintergreen, or seemed to be irrelevant to Vincent's lifetime. Those biochemical factors which we have
discussed are at least reasonable candidates for causing confusion or distortion in color perception. Some biophysical aspects are also worth mentioning.

Biophysical interference

Color information received by the eye requires the functional and structural
integrity of a great number of specialized sensory cells, interconnecting
nervous pathways, and associated centers of the brain, before it is registered
as useful knowledge. Interference at virtually any stage in the overall process
can result in aberrancies in perception. Physical aspects include the
development of opacities (cataracts) in the lens, changes in the relative
amounts of the different color sensing elements (cones) in the retina,
pathology of the optic nerve, and the mechanical effect of tumors imping-
ing on nerves or specific lobes of the brain.

Cortical cataracts are grey and function as a neutral filter. On the other
hand, so-called nuclear cataracts have a yellowish pigmentation and function as an endogenous yellow filter. A degree of compensation attends and
the xanthopsia of nuclear cataracts depends largely on the speed with which
they develop. A painter who experiences a relatively rapid onset of cataracts
becomes less distinct in his landscapes and concomitantly tends to avoid the
blue pigments; his palette is dominated by the reds. When a cataract is
removed the patient suddenly sees the blue sky as well as in his youth and,
for a short time at least while the brain adjusts, there is an overindulgence
in the blue pigments. The classical example is provided by Claude Monet.[66]

The high acuity of the healthy eye resides in the ability to focus objects
into one small region in the center of the retina called the macula. (The
remainder of the retina provides peripheral vision of low resolution as well
as night vision.) The macula is also responsible for color vision, which is
achieved through the agency of S (blue) [maximally sensitive at 440 nm],
M (green) [535 nm] and L (red) [565 nm] cones. It is well accepted that
the L- and M-receptors are more numerous than S-receptors, and the S-cone
pathway is said to have a more limited response range,[67] as well as being
relatively more fragile.[68] Thus diseases such as diabetes and retinitis pigmentosa which lead to macular degeneration are all attended by a decrease

in the sensitivity of the S (blue) cone pathway. A diabetic retinopathy supposedly affected the palette in Cézanne's later work.[3]

Glaucoma is a disease in which the intraocular pressure is raised. The initial effects are on peripheral vision which is fogged, and bright lights are ringed with haloes. If this elevation is prolonged, the pressure eventually destroys the retina and the optic nerve. Although central vision is the last to go, controlled studies with patients suffering from chronic glaucoma also indicate defects in blue-yellow discrimination,[69] supposedly due to the higher sensitivity of the blue cone receptors.[68]

Any damage to the optic nerve will result in a disturbance of conductivity which can modify or, if the insult is sufficiently severe, completely destroy color sensation. With inflammatory, toxic, or mechanical (tumor) disturbances to nerve fibers the response to red seems to be the most vulnerable.[70] Lesions at the anterior inferior part of the occipital lobe are also associated with chromatopsias although primary visual function is preserved.[71] A well-documented case for this type of aberrancy concerned two red paintings by the English artist Percy Wyndham Lewis (1882–1957). According to Conway,[72] the shock of the artist on finding himself deserted by his sense of a primary color caused an overengagement with red pigments. In the first portrait of the artist's wife she was seated before an open fire; in the second she was clothed in his own red dressing gown. Both paintings are remarkable for their pervading rich red coloration. An unusually large pituitary tumor was found at autopsy. The chromatopsia was apparently caused by the tumor pressing upon and damaging the optic chiasm.

There is no direct evidence to suggest that Vincent suffered from any of these biophysical abnormalities. An isolated claim for glaucoma has not attracted many adherents. Advanced cataracts would surely have been discovered. Both conditions are essentially unremitting if not treated and do not fit the episodic incidence of high yellow paintings.

We turn now to *artistic preference*. The key distinction in our operational definition is that the artist views the motif accurately but elects to depart from realistic depiction of colors on the canvas. That the artist's choice of pigments be conscious, accurate, and uninfluenced by external factors is central to this argument.

Vincent's preference

In 1886 van Gogh declared that, "true drawing is modelling with color" (letter 459a). Later, in *La Berceuse (Woman Rocking the Cradle)*, F 504, Arles, 1889, he strove for facial modelling by "naturally broken tones" upon a "complexion [of] chrome yellow" and hoped, above all, to paint a "lullaby in colors" (letter 571a). Here, and elsewhere, the artist invokes a deeper emotional language as well as a new visual technique; the preoccupation is with color. And elsewhere there is ample evidence from van Gogh himself to indicate that he was prepared to occasionally depart from "exact" color (letter 533). An illustrative example may be his rendering of impressions of *The Night Café*, F 463, "to express the terrible passions of humanity by means of red and green" (letter 533). The disparate reds, greens, orange, and yellows achieve, "an atmosphere like a devil's furnace" (letter 534), and compare with Gauguin's *In an Arles Café*, 1888, Pushkin Museum, Moscow, for the application of a cooler palette to render the same room.

His preference can be gauged from letters (1887–1890) wherein he mentions the yellow of his surroundings more than any other color. Yellow also had a special symbolism for Vincent: *The Wheatfield behind Saint Paul's Hospital with a Reaper*, F 617, 1889, a work predominantly in yellow, was described as, "a vague figure fighting like a devil ... the image of death ... [and] humanity might be the wheat he is reaping" (letter 604). But, most important for Vincent, "there is nothing sad in [this] death, [because] it goes its way in broad daylight ... with a light of pure gold" (letter 604). This is reminiscent of Goethe's conviction that the colors of yellow, orange, and cinnabar (vermilion) evoke quick, lively, aspiring feelings and that yellow is closest to the light of the sun.[73] Later van Gogh hesitated over the color rendition of this picture, but "preferred the canvas done from nature [over a later copy] ... [because it] makes me recall the furnace of summer ... *it is not so exaggerated after all*" (my italics) (letter 608).

Conclusion

We obviated natural ageing of the lens because of Vincent's youth and short career span. Digitalis intoxication has sufficient medical underpinning but lacks usage documentation and does not fit the multi-regional aspect of the yellow palette. The physiological case for overdosage on santonin is as good or better than that for digitalis, but the evidence is still circumstantial, viz. Raspail's book and the terpene connection. Sunstroke does not fit the multiplicity of motifs or locations for the yellow paintings. There was no evidence of cataracts or glaucoma in Vincent's eyes. A minority of the paintings may have been directly influenced by absinthe-induced hallucinations. On the other hand, irreversible brain damage from absinthism is not indicated because it does not fit the episodic nature of the yellow paintings.

Just as one feels ready to embrace "artist's choice," and reject all other hypotheses for the reasons given, we are reminded of Vincent telling his brother, about his early months in Arles, "Now if I recover, I must begin again, and I shall not again reach the heights to which sickness partially led me" (letter 570). "Dr Rey says that … I kept myself going on coffee and alcohol … it is true that to attain the high yellow note that I attained last summer, I really had to be pretty well keyed up" (letter 581). And with reference to Albert Aurier's glowing review in 1890 he wrote to his sister Wil, "all the impressionists are like that [the features attributed to Vincent by Aurier], are under the same influence, and we are all of us more or less neurotic. This renders us very sensitive to colors and their particular language, the effects of complementary colors, of their contrasts and harmony" (letter W20). How difficult it is to completely dismiss the influence of illness and external effects on his color perception! Nonetheless, artistic preference remains the best working hypothesis to explain the yellow dominance in Vincent van Gogh's palette.[1]

References and notes

1. Arnold WN & Loftus LS. 1991. Xanthopsia and van Gogh's yellow palette. *Eye* **5**: 503–510.
2. Arnheim R. 1986. *New Essays on the Psychology of Art.* p 160 & p 209. Berkeley, Los Angeles, London: University of California Press.
3. Trevor-Roper PD. 1959. The influence of eye disease on pictorial art. *Proceedings of the Royal Society of Medicine* **52**: 721–744.
4. Medawar P. 1986. *Memoir of a Thinking Radish.* Oxford & New York: Oxford University Press.
5. Mills L. 1936. Peripheral vision in art. *Archives of Ophthalmology* **16**: 208–219.
6. Coren S, Porac C & Ward LM. 1978. *Sensation and Perception.* New York: Academic Press.
7. Siegel MH, Dimmick FL. 1962. Discrimination of color. II. sensitivity as a function of spectral wavelength, 510 to 630 nm. *Journal of the Optical Society of America* **52**: 1071–1074.
8. Boynton RM. 1956. Rapid chromatic adaptation and the sensitivity functions of human color vision. *Journal of the Optical Society of America* **46**: 172–179.
9. Land EH. 1959. Experiments in color vision. *Scientific American* **200**: 84–99.
10. Boettner EA & Wolter JR. 1962. Transmission of the ocular media. *Investigative Ophthalmology and Visual Science* **1**: 776–783.
11. Pokorny J, Smith VC & Lutze M. 1987. Aging of the human lens. *Applied Optics* **26**: 1437–1440.
12. Hochberg J. 1971. Perception: I. color and shape. pp 395–427, in: *Woodworth & Schlosberg's Experimental Psychology.* 3rd ed. JW Kling & LA Riggs (eds.) New York: Holt, Rheinhardt & Winston Inc.
13. Goethe JW. 1970. *The Theory of Colours.* p 307 (translated by CL Eastlake) Cambridge: MIT Press.
14. de la Faille JB. 1970. *The Works of Vincent van Gogh: His Paintings and Drawings.* New York: Reynal & Co.

15. Tralbaut ME. 1981. *Vincent van Gogh.* New York: The Alpine Fine Arts Collection Ltd.
16. Pickvance R. 1984. *Van Gogh in Arles.* New York: Harry N. Abrams Inc.
17. Pickvance R. 1986. *Van Gogh in Saint Rémy and Auvers.* New York: Harry N. Abrams Inc.
18. Stein S. 1986. *Van Gogh: A Retrospective.* New York: Park Lane.
19. Barrielle JF. 1984. *The Life and Work of Vincent van Gogh.* Secaucus, New Jersey: Chartwell Books Inc.
20. Zurcher B. 1985. *Vincent Van Gogh: Art, Life and Letters.* New York: Rizzoli.
21. Hulsker J. 1980. *The Complete van Gogh: Paintings, Drawings, Sketches.* New York: Harry N. Abrams Inc.
22. Bonafoux P. 1989. *Van Gogh Self Portraits.* New York: Tabard Press.
23. Walsh FB & Hoyt WF. 1969. vol 3, pp 2541–2716, in: *Clinical Neuro-Ophthalmology.* 3rd ed. Baltimore: Williams & Wilkins Co.
24. Withering W. 1785. *An Account of the Foxglove and Some of its Medical Uses: with Practical Remarks on Dropsy and other Diseases.* Birmingham: GGJ & J Robinson. (This book was reproduced under the same general title, with annotations by Aronson JK. 1985. London: Oxford University Press.)
25. Hanzlik PJ. 1925. Jan Evangelista Purkyne (Purkinje) on disturbances of the vision by digitalis, one hundred years ago. *Journal of the American Medical Association* **84**: 2024–2025.
26. Duke-Elder S. 1971. *System of Ophthalmology.* vol 12, p 552. St. Louis: CV Mosby Co.
27. Sprague HR, White PD & Kellogg JF. 1925. Disturbances of vision due to digitalis: review of the literature and report of cases. *Journal of the American Medical Association* **85**: 716–720.
28. Lely AH & van Enter CHJ. 1970. Large-scale digitoxin intoxication. *British Medical Journal* **3**: 737–740.
29. Duke-Elder S. 1972. *System of Ophthalmology.* vol 14. St. Louis: CV Mosby Co.
30. Gibson HC, Smith DM & Alpern M. 1965. Π_5 specificity in digitoxin toxicity. *Archives of Ophthalmology* **74**: 154–158.
31. Unger L. 1958. Chromatopie nach Digitalis. *Ophthalmologica* **136**: 326–332.
32. Robertson DM, Hollenhorst RW & Callahan JA. 1966. Ocular manifestations of digitalis toxicity. *Archives of Ophthalmology* **76**: 640–645.
33. Lee TC. 1981. Van Gogh's vision: Digitalis intoxication? *Journal of the American Medical Association* **245**: 727–729.
34. Rentchnick P. 1987. Pathographie, van Gogh. *Médecine et Hygiene* **45**: 1750–1761.
35. Lanthony P. 1988. La xanthopsie de van Gogh. *Médecine et Hygiene* **46**: 2466–2472.
36. Ravin JG. 1981. Van Gogh's Illness. *Ohio State Medical Journal* **77**: 699–702.
37. Weiner D. 1968. *Raspail, Scientist and Reformer.* New York & London: Columbia University Press.
38. Raspail FV. 1886. *Manuel Annuaire de la Santé pour 1886.* Paris: Chez l'Editeur. (This is the edition available to me, see chapter 7 for further information).
39. Nordenfalk C. 1947. Van Gogh and literature. *Journal of the Warburg and Cortauld Institutes* **10**: 132–147.
40. Wood GB & Bache F. 1886. *The Dispensatory of the United States of America.* 15th ed., p 1270. Philadelphia: JB Lippincott & Co.
41. Simonsen JL. 1949. *The Terpenes.* vol 2. Cambridge: Cambridge University Press.
42. Wood CA. 1919. *The American Encyclopedia and Dictionary of Ophthalmology.* vol 17, p 12861. Chicago: Cleveland Press.

43. Marshall W. 1927. A study of santonin xanthopsia. *Journal of Pharmacology and Experimental Therapeutics* **30**: 361–388.
44. Grant WM. 1974. *Toxicology of the Eye.* 2nd ed. Springfield, Illinois: Charles C. Thomas.
45. Cookson HA & Stock CJH. 1940. Santonin poisoning, fatal case. *Lancet* vol 2 for 1940: 745.
46. Oldham RR, Wang YM, Van Eys J & Carter JP. 1971. Hemolytic crisis in a patient treated with santonin: possible santonin poisoning. *Southern Medical Journal* **64**: 480–482.
47. Arnold WN. 1988. Vincent van Gogh and the thujone connection. *Journal of the American Medical Association* **260**: 3042–3044.
48. Monroe RR. 1978. The episodic psychoses of Vincent van Gogh. *Journal of Nervous and Mental Disorders* **166**: 480–488.
49. Schmidt H. 1915. L'Absinthe l'alienation mentale et la criminalite. Rapport fait au nom de la commission d'hygiène publique de la chambre des députés. *Annals d'Hygiene Publique et Medecine Legale* 4th series, **23**: 121–133.
50. Arnold WN. 1989. Absinthe. *Scientific American* **260**: 112–117.
51. Magnan V. 1874. On the comparative action of alcohol and absinthe. *Lancet* vol 2 for 1874: 410–412.
52. Sollmann T. 1948. *A Manual of Pharmacology and Its Applications to Therapeutics and Toxicology.* 7th ed. Philadelphia: WB Saunders Co.
53. Kalinowsky LB, Hippius H & Klein HE. 1982. *Biological Treatments in Psychiatry.* chapter 3, pp 217–221. New York: Grune & Stratton.
54. Dean SR. 1940. Studies in convulsant therapy. IV. The effects of metrazol (pentamethylenetetrazol) on the eye. *Archives of Ophthalmology* **24**: 316–325.
55. Friedman E. 1937. Irritative therapy of schizophrenia. *New York State Journal of Medicine* **37**: 1813–1821.
56. Muskens LJJ. 1928. *Epilepsy.* p 274. New York: William Wood and Company.
57. Hilbert R. 1913. Zur Kenntnis der genuinen Chromatopien. *Klinische Monatsblätter für Augenheilkunde* **51**: 494–497.
58. Arnold WN, Dalton TP, Loftus LS & Conan PA. 1991. A search for santonin in *Artemisia pontica,* the other wormwood of old absinthe. *Journal of Chemical Education* **68**: 27–28.
59. Beer J. 1935. *Essai sur les Rapports de l'Art et de la Maladie de Vincent van Gogh.* Medical Thesis, University of Strasbourg.
60. Grey R. 1924. *Vincent van Gogh.* Rome: Valori Plastici Publishers.
61. Beer J. 1950. Van Gogh: diagnosis of the tragedy. *Art News Annual* **19**, special number: 82–90
62. Lubin AJ. 1987. *Stranger on the Earth: A Psychological Biography of Vincent van Gogh.* New York: Henry Holt & Co.
63. Fabricant J. 1958. Heat stroke. *US Armed Forces Medical Journal* **9**: 1106–1118.
64. Satterthwaite TE. 1886. Heat-stroke. vol 3, p 600, in: *A Reference Handbook of the Medical Sciences.* AH Buck (ed.). New York: William Wood & Co.
65. Carroll FD. 1945. Visual symptoms caused by digitalis. *American Journal of Ophthalmology* **28**: 373–376.
66. Ravin JG. 1985. Monet's cataracts. *Journal of the American Medical Association* **254**: 394–399.

67. Hood DC, Benimoff NI & Greenstein VC. 1984. The response range of the blue-cone pathways: a source of vulnerability to disease. *Investigative Ophthalmology & Visual Science* **25**: 864–867.

68. Greenstein VC, Hood DC, Ritch R, Steinberger D & Carr RE. 1989. S(blue) cone pathway vulnerability in retinitis pigmentosa, diabetes and glaucoma. *Investigative Ophthalmology & Visual Science* **30**: 1732–1737.

69. Foulds WS, Chisholm IA & Bronte-Stewart JM. 1973. Effects of raised intra-ocular pressure on hue discrimination. Colour Vision Deficiencies II. International Symposium, Edinburgh (1973), published in: *Modern Problems in Ophthalmology* **13**: 328–334 (1974).

70. Huber A. 1971. *Eye Symptoms in Brain Tumors*. 2nd ed. Saint Louis: C. V. Mosby Company.

71. Meadows JC. 1974. Disturbed perception of colours associated with localized cerebral lesions. *Brain* **97**: 615–632.

72. Conway JF. 1988. An early effect of Wyndham Lewis' pituitary tumour on his art: an inquiry prompted by a note in "The World through Blunted Sight." *Eye* **2**: 677–681.

73. Goethe JW. 1970. *The Theory of Colours*. pp 304–308. (translated by CL Eastlake) Cambridge: MIT Press.

Chapter 9
The Ear-Cutting Affair

Figure 9.1. Fernando Gallego (b. circa 1440) and assistants, *The Betrayal of Christ*, 155 × 110.5 cm (61 × 43.5″), oil on canvas applied to wood, panel 16, The Retablo of Ciudad Rodrigo. Collection of The University of Arizona Museum of Art, Tucson, gift of Samuel H. Kress Foundation.

The subject is taken from the Bible. "Then Simon Peter having a sword drew it, and smote the high priest's servant, and cut off his right ear. The servant's name was Malchus." (The Gospel according to St. John, Chapter 18, verse 10, written between A.D. 85 and 90).

*I shall always believe in the art that is to be
created in the tropics ... but personally I am too
old and, especially if I have a papier mâché ear
put on, too jerry-built to go there.*
Vincent to Theo, letter 574, from Arles,
January 28, 1889.

Paul Gauguin wrote to Theo van Gogh, in the latter half of December, 1888, expressing incompatibility of temper with Vincent, requesting funds for the return trip to Paris, and apologizing for his impending leave-taking.[1] The quarrels ceased for the moment and Gauguin then asked Theo to regard his previous letter "as a bad dream." Meanwhile, Vincent waited for Gauguin to make a more definite decision with, in his words, "absolute serenity" (letter 565, December 23). Gauguin's rendition[2] of the days before Christmas is not regarded as completely reliable, but suggests that Vincent accosted him in the street with an open razor. Gauguin spent the night in a hotel. Vincent supposedly returned to the Yellow House on his own, turned the same instrument on himself, and committed the infamous self-mutilation of the left ear.

Johanna van Gogh-Bonger[1] recalled that "a telegram arrived from Gauguin that called Theo to Arles. Vincent, in a state of terrible excitement and in a high fever had cut off a piece of his own ear and taken it as a present to a woman in a brothel. There had been a violent scene; Roulin the postman managed to get him home [other accounts have Vincent returning to the Yellow House alone], but the police intervened, found Vincent bleeding and unconscious in bed, and sent him to the hospital." In a letter to Gauguin (appended to letter 566, January 1, 1889) Vincent clarified the vital role of Roulin who "has been truly kind to me, it was he who had the presence of mind to make me come to that place [the Hospital, the following morning] before the others were convinced." From all of this, as well as Gauguin's own account,[2] it is clear that Vincent's erstwhile companion in art quickly departed both the domicile and the town. It was Vincent's charwoman and Roulin who took care of the Yellow House and put everything back in order (letter 566).

The ear-cutting affair is obviously important from several perspectives; it was part of Vincent's first crisis in Arles, it precipitated institutional attention, it provided incontrovertible evidence of aberrant behavior, it spawned accusations of madness, and the stigma frightened and plagued him. It also constitutes one of the best known, albeit morbid, aspects of the van Gogh persona. Accordingly, there has been much popular interest in the cause of the act, and a plethora of hypotheses surround this bizarre incident.

Vincent's recollections and comments

Analysis is made exceedingly difficult by the victim's claim that he had no recollection of the event (letters 576, 587, W11). Some remarks in letter 573, January 23, 1889, suggest a subconscious attempt to obtain relief from torment in the hours or days after the event, "During my illness I saw again every room in the house at Zundert [my birthplace], every path, every plant in the garden, the views of the fields outside, the neighbors, the graveyard, the church, our kitchen garden at the back – down to a magpie's nest in a tall acacia in the graveyard." There is but one tantalizing indication of more recall of the bizarre incident: "I say no more about it, since I had better not try to go over all that passed through my head then" (letter 573). He never developed this aspect and in a later letter concluded with a denial: "for the most part I have completely lost the recollection of those days in question, and can reconstruct none of it" (letter 587). His nightmares were also poorly documented but included "fits of dizziness" (letter W4), "an absurd religious turn" (letter 605), and "perverted and frightful ideas about religion" (letter 607).

It is worth mentioning that the other incident of self-mutilation – "Let me see her for as long as I can keep my hand in the flame," occasioned by an attempt to win the heart of his widowed cousin, was also poorly remembered; "I think they blew out the lamp" (letter 193). This wretched event occurred in Amsterdam, at the end of 1882.

Vincent made only one passing reference to his need for "a papier mâché ear" (letter 574), which I take to be humor. Four months later, while engaged in a discussion of symptoms, he wrote, "Most epileptics bite their tongue and injure themselves. [Dr.] Rey told me that he had seen a case where someone had mutilated his own ear as I did" (letter 592). There is no data to support this euphemism. Favazza[3] is correct in claiming that self-mutilation of the ear is very rare in the medical literature.

Missing and sacrificed ears

The English author Daniel Defoe (1659–1731) lacked external ears from birth.[4] The French litterateur Pierre Alphonse Guys[5] (1755–1812) and the post-impressionist Henri Rousseau[4] (1844–1910) were born with only one ear a piece. Several cases of congenitally absent or deficient aural appendages have been documented.[5] Absence of the external auditory meatus (the ear-hole) is encountered more frequently;[5] the report by Munro[6] is such an example, possibly made more notable by the fact that an experienced nurse bathed the child daily for *five* weeks before the abnormality was noticed.

Finally, we should mention the nine inmates of Pentridge Prison, Melbourne, Australia, who each cut off an ear, or part of an ear; they are known locally as the van Gogh club.[7]

The morbid details of Vincent's ear

Tralbaut[8] has a diagram of the left ear and the part severed. On the same page we find the newspaper report about Vincent's enjoining the prostitute Rachael to "guard this object [the excised piece] carefully." The tissue supposedly ended up in the hospital at Arles, was kept in alcohol for a year, and then discarded.[9]

Yasuda[10] gave the most plausible description of the self-mutilation process, "van Gogh [probably] pulled up the lobule of his [left] ear with his left hand and then slashed the ear upward with the razor held at the lower attachment of his ear. While the razor could cut soft tissues, it struck the auricular cartilage and the [knife] edge turned outward and as a result, only the lower half of his ear was slashed off." The fact that Vincent was right-handed would be sufficient to explain the attack on the *left* ear. The imagined scene was rendered by Rodolophe Pissarro, using Vincent van Gogh's painting of his bedroom as a model background.[11] Vincent painted two mirror-images in Arles depicting the bandaged left ear. In the three subsequent self-portraits from St. Rémy, also mirror-images, the subject faces the other direction and the pictures show the unaffected right ear.

Hypotheses

Given that Vincent had neither plan nor recollection of the ear-cutting affair, many of the working hypotheses that have appeared in the literature seem contrived and inappropriate. According to Lubin,[12,13] van Gogh's ear incident, more than anything else, "provided a powerful stimulus for the naïve but popular notion that great artists are apt to be insane." Runyan[14] briefly summarized 13 suggestions regarding van Gogh's ear but did not reach a conclusion, supposedly because he was more concerned with the theory of alternative explanations and used van Gogh's act as an exemplar. I have taken the approach that the circumstances surrounding the affair encourage only two types of explanation, i.e. an accident or an acting out of unconscious themes.

248

Accident

In the painting by Gauguin, *Van Gogh Painting Sunflowers*, which can be dated to the second half of September 1888,[15] Vincent is depicted with a full red beard. Shortly thereafter, Vincent wrote,"I have just painted my own portrait, in my own ashen coloring" (letter 540); referring to portrait number 40 in Bonafaux[15] which depicts a clean-shaven face. A self-portrait from November–December, inscribed "to friend Laval," shows stubble at most. (Incidentally, this is the last depiction of the intact left ear.) In the two bandaged images immediately after the ear-cutting, from January 1889, the face is also clean-shaven. Thus Vincent was shaving at this time on a more or less regular basis and therefore during the Christmas crisis *it is possible* that he was thus engaged, suffered a seizure, and cut himself. Alternatively, if hallucinating, he might have struck out at something imagined on his face. (We can further speculate that upon recovery the victim discovers the lopped off ear lobe, wraps it up, and takes it to the prostitute as a gift.) This simple explanation has neither more nor less proof than any other hypothesis in the literature; if true, what a lot of ink has been wasted!

Acting out during a medical crisis

This type of explanation depends upon the victim hallucinating, or at least being in a confused state, irrespective of whether he is suffering from a toxic psychosis or some other ailment. As a working hypothesis, we suppose that a medical crisis triggers the release of ideas embedded in the subconscious, which are then acted out. In our exploration of possible influences we will start with two examples from books. Vincent was quite conversant with these sources and he provided ample evidence of being deeply impressed by their themes.

The Frère Archangias model

And Jeanbernat was in fact entering the cemetery. He walked straight to the group around the grave. His step was as sprightly as ever; he was still so nimble that he made no noise. He stood behind Brother Archangias and seemed for an instant to be gazing intently at the back of his neck. Then, as Father Mouret was finishing the prayers, he calmly pulled a knife from his pocket, opened it, and chopped off the friar's ear.
No one had time to interfere. The friar screamed.

"The left one's for later," Jeanbernat said peacefully as he threw the ear to the ground.

This scene is the penultimate climax to Zola's novel, *The Sin of Father Mouret*.[16] The story is set in the south of France, in the fictitious village of Artaud, whose inhabitants are inbred and seemingly outside the laws of human conventions. Serge Mouret is the village priest who suffers a religious ecstasy and mental collapse. His cousin, Dr. Pascal Rougon, initiates rehabilitation nearby, in an overgrown, walled park, *Le Paradou*. (There is a small town of this name, not far from Arles and St. Rémy.) Mouret is placed under the nursing care of Albine, the 15-year-old daughter of the local philosopher, Jeanbernat. Serge frolics in Paradou with the young girl, impregnates her (the "sin" of the title), but is "rescued" by the village teacher, Brother Archangias. The boisterous Archangias has a proclivity for pulling children's ears, not least of which those of an altar boy with red hair *named Vincent*. Mouret suffers hallucinations but resists Albine, who then commits suicide. At her burial Jeanbernat takes his revenge on Archangias. The coffin is finally lowered and, as one of the peasants is throwing in the first spadeful of dirt, Father Mouret's sister yells that the cow has had a calf.

Émile Zola, the most celebrated member of the ultra-realistic French writers, the so-called naturalists, wrote *The Sin of Father Mouret* in 1875 as the fifth work in the Rougon-Macquart cycle. He recast Genesis into a fictional tale of man in nature. The following quote exemplifies the style, and has an interesting reference to wormwood:

And, despite themselves, yielding to some force driving them [Serge and Albine], they went around a rock and climbed up to a plateau, where the intoxication of the bright sun was awaiting them. They no longer found the happy languor of aromatic plants, the musk of thyme, the incense of lavender. They were crushing stinkweeds; the bitter alcoholic breath of wormwood; the rotting-flesh smell of rue; burning valerian dripping its aphrodisiac sweat; mandrake, hemlock, setterwort, and belladonna sent dizziness to their temples, weakened them, made them tremble in one another's arms, their hearts on their lips.[16]

Zola's "bitter alcoholic breath of wormwood" is not quite right, for the plant does not smell of alcohol, he was surely thinking more of absinthe.

Vincent van Gogh first read this novel in The Hague. On August 19, 1882 he wrote, "I have read most of Zola's *La Faute de l'Abbé Mouret* and *Son Excellence Eugène Rougon*, both beautiful. I think Pascal Rougon, the doctor who appears in his series of books, but always in the background, a noble figure. He really proves that no matter how degenerate a race may be, it is always possible for energy and will power to conquer fate. In his profession he found a force stronger than the temptation he had inherited from his family; instead of surrendering to his natural instincts, he followed a clear, straight path, and did not slide into the wretched muddle in which all the Rougons perished. He and Madame François of *Le Ventre de Paris* are to me the most sympathetic figures" (letter 226). A couple of months later he wrote to Van Rappard, "When I was ill [in The Hague] and afterward I read Zola's books with great admiration" (letter R13).

The next year Vincent compared one of Theo's escapades of the heart in Paris with the events in *La Paradou*, and suggested that Theo and his girlfriend might be a good model for a proposed painting (letter 286). In the next letter he also thought of "someday ... [attacking] a Paradou subject" (letter 287).

In Arles, before the ear-cutting incident, he was reminded of the overgrown park, "I have come back from a day in Mont Majour ... We explored the old garden together ... If it had been bigger it would have made me think of Zola's Paradou" (letter 506). Earlier, Vincent declared that Zola had deficiencies as an art critic (letters 418, R38), but he loved Zola's comment, "in the picture [work of art] I seek, I love, the man – the artist" (letter 418). He contrasted his father's lack of awareness of "the soul of modern civilization" with Zola's understanding, great simplicity, and truth (letter 339a). He recommended his brother Theo (letters 212, 219, 418), sister Wil (letters W1, W14), and friend Bernard (letter B13) to not only read Zola but urged that "[this] is good for one, and makes things clear" (letter 219). He was so enthused that he wanted "to read everything by Zola" (letter 212). And later, while in the St. Rémy asylum, he repeated his admiration for Zola (letter 607).

Kodera,[17] in his recent book *Vincent van Gogh: Christianity versus Nature*, devotes a whole chapter to *La Faute de l'Abbé Mouret* but is more than cautious about comparing the ear-cutting episode in the book with the van Gogh episode. Not so Byatt,[18] who makes a very reasonable case for it in his review of Kodera's book. There can be little doubt that Vincent was as fond of Zola as he was of the Bible even if he confessed that "A Paradou is beautiful, but Gethsemane is even more beautiful" (letter 319).

The Garden of Gethsemane model

*When they that were about him saw what
would follow, they said unto him, Lord, shall
we smite with the sword?
And one of them smote the servant of the high
priest, and cut off his right ear.
And Jesus answered and said, Suffer ye thus
far. And he touched his ear, and healed him.*
Luke 22: 49–51.

Matthew, Mark, and John relate similar events; John adds that it was Peter who wielded the sword and Malchus who was struck. Luke, in keeping with his first profession as physician, is the only one to include the restoration of the ear.

Six months before the ear-cutting affair, Vincent wrote to Bernard:

the study of Christ inevitably calls it [the artistic neurosis] forth, especially in my case where it is complicated by … innumerable pipes [of tobacco]. The Bible is Christ, for the Old Testament leads up to this culminating point. St. Paul and the evangelists dwell on the other slope of the sacred mountain [Mount of Olives]. … But the consolation of that saddening Bible which arouses our despair and our indignation … is Christ. … of all the philosophers … [He] has affirmed … eternal life, the infinity of time, the nothingness of death, the necessity and the *raison d'etre* [justification] of serenity and devotion. He lived serenely, as a greater artist than all other artists, despising marble and clay as well as color, working in living flesh. … He made … living men. … This great artist did not write books either; surely Christian literature as a whole would have filled him with indignation, and very rare in it are literary products that would find favor in discerning eyes beside Luke's Gospel or Paul's Epistles … [His] spoken words … are one of the highest summits … reached by art … a pure creative power. These considerations … raise us above art itself. They make us see the art of creating life, the art of being immortal and alive at the same time. They are connected with painting. The patron saint of painters, St. Luke – physician, painter, evangelist – whose symbol is, alas, nothing but an ox, is there to give us hope (letter B8).

Earlier, Vincent quoted Gustave Doré on the definition of a painter and the arguments about natural artistic gifts, "J'ai la patience d'un boeuf" [I have the patience of an ox] (letter 336). Luke was said to have been a preacher in Egypt and Greece after the death of Paul, who had described him as the "beloved physician" (Colossians 4:14). Also, he was popularly supposed to be a painter and became patron saint of artists.

The theme is further developed in his next letter to Bernard: I cannot help thinking that you may well be surprised to see how little I like the Bible, although I have often tried to study it a little. ... But Christ, I repeat, is more of an artist than the artists; he works in the living spirit and the living flesh, he makes men instead of statues. And then ... I feel only too well that I am an ox – being a painter – I, who admire the bull, [the lion?], the eagle, the man, with a veneration that will prevent me from being ambitious (letter B 9). (The early Christian church, which made much use of symbolic imagery, sometimes represented the evangelists as four winged creatures; Matthew as a man [angel], Mark as a lion, Luke an ox, John an eagle. See Ezekiel 1:5–14 where the prophet tells of the strange vision of the four beasts.)

Later in the year, probably the first half of October but before Gauguin arrived, Vincent wrote to his sister, "I think of Monticelli terribly often here. ... He died at Marseilles in rather sad circumstances, and probably after passing through a regular Gethsemane. ... I am continuing his work here, as if I were his son or his brother" (letter W8). Also, in a letter to Theo, we find simulation, "I have bought 12 chairs [12 apostles?]" (letter 534); unfulfilled veneration, "For the second time I have scraped off a study of Christ with the angel in the Garden of Olives" (letter 540); and a key position for Theo in the proposed artists' colony, "So you [Theo] will be one of the first, or the first dealer-apostle" (letter 544) – all with patently biblical references. All of the foregoing remarks can be compared with Vincent's later statement from St. Rémy, "I get perverted and frightful ideas about religion" (letter 607). A plausible background theme and a model for his self-mutilation can certainly be gleaned from the Bible and the Garden of Gethsemane. If that is the *Petrine* hypothesis we will now consider the so-called[18] *Taurine* hypothesis.

The matador and the bull's ear model

This hypothesis is attributed to J. Oliver, native of St. Rémy, who wrote to V. W. van Gogh, Vincent's nephew[*] in 1951. The substance of that communication, which is recorded as item A13 in *The Complete Letters,* is as follows. A

---

[*] Dr. Albert Lubin told me, in October 1990, that V. W. van Gogh, *the engineer,* was also something of an amateur psychiatrist. This may help to explain the great number of psychological biographies of Vincent van Gogh that were endorsed by his nephew.

matador, whose skill has been recognized by audience acclaim, is awarded the bull's ear by the president of the arena. An assistant cuts off the ear, hands it to the matador, who then tosses it to a female spectator. The correlation with Vincent's self-mutilation apparently rests on his simultaneously playing the roles of both bull and matador's assistant. Rachael, the prostitute, is the lucky woman. Lubin[13] called the hypothesis a refreshingly new interpretation; he and other psychoanalysts have no trouble accommodating this duality and have also invoked sundry corollaries based on the ear as a sexual symbol, the male fear of castration, and so forth. For those who need a stimulus to the imagination along these lines, probably the best place to start is either Lubin[13] or Nagera[19] wherein even the objects in Vincent's painting are richly endowed with sexual symbolism. For example, the billiard cue on the table of *The Night Café* is equated with a penis and the curtained exit with a vagina.

Bullfights occurred every Sunday during the summer at Arles (letter 498a); we know that Vincent witnessed a couple. He recorded his thoughts: "Yesterday I saw another bullfight, where five men played the bull with darts and cockades. One toreador crushed one of his balls jumping the barricade. [Author's note: the possible anatomical relationships are increasing!] He was a fair man with grey eyes, plenty of *sang-froid* [coolness of blood, mind]; people said he will be ill long enough. ... The arenas are a fine sight when there's sunshine and a crowd" (letter 474). And to Bernard he wrote, "I have seen bullfights in the arena, or rather sham fights, seeing that the bulls were numerous but there was nobody to fight them" (letter B3). There is even some argument in the literature[20] as to whether the bulls were actually killed – as in the Spanish custom – in Arles during van Gogh's time, all of which does some damage to the taurine hypothesis. Vincent painted just one picture of the bullfight, *A View of the Arena in Arles*, F 548, which has images of the Roulins in the crowd. The artist may have been more interested in the festive occasion than the cruelty of the bullfight.

Animosities toward father and substitute figures

Schnier[21] is one of the champions of this hypothesis. He feels that Vincent saw some of the father figure in the older, more-experienced (in art and women) Paul Gauguin, whose arguments in art history set off a powder keg of old animosities. Basic to this theory is the supposed relationship between the ear lobe and sexual organs, plus an Oedipus complex in Vincent (a need to do away with his father to be closer to his mother), and of course the good old general fear of castration. Also, we are asked to bear in mind that self-mutilation so offends man's instinct for self-preservation that it can only occur as an indirect punishment of another.

According to Schnier, Gauguin (as the father) was threatened by Vincent, but escaped. Van Gogh turned on himself, by now identifying with his father, whom he punished by removal of the sexual symbol (the ear lobe). This working hypothesis assumes that Vincent was committing an act of violence to his father and punishing himself *at the same time.* The severest possible punishment, comparable almost to death itself, would be the loss of his masculinity. By castration of the ear? And so it goes; the fact that Vincent's father died three years earlier is accommodated by Schnier as unresolved tensions in Vincent plus insufficient grieving. To make things even more complicated, a strong bisexual predisposition is supposed to be indicated by Vincent's interest in communal art, attempts to emulate his father as a preacher, his relationship with Gauguin, and his closeness to Theo. Imaginative as this theory may be, one wonders how far it advances the field. A critical essay on some of the general problems which emerge in subjecting this type of psychoanalysis to objective assessment has been offered by Storr.[22]

Other speculations

The paper[12] and book[13] by Lubin and the references therein should be consulted for extensive lists of possible causes and psychiatric analyses. Manifold stresses have been invoked as precipitants of the ear-cutting affair. Some of these are Theo's engagement to Johanna; Vincent's competition with Johanna as well as his mother for Theo's attention during the upcoming Christmas celebration; and strained relationships between Vincent and Gauguin. Some commentators have supposed that Vincent, through self-mutilation, was endeavoring to gain sympathy from his mother, Theo, or the Roulin family. Lubin[13] and Runyan[14] entertained the concept that the much publicized Jack the Ripper case in London, involving a serial killer who had mutilated the bodies of murdered prostitutes, may have been influential. This hypothesis apparently views Vincent's act as a masochistic twist to a sadistic news event.

Concluding remarks

Vincent's ear-cutting affair at the end of 1880 should also be considered in the broader aspect of his suicide, which occurred a year and a half later, and is the subject of the next chapter. Important here is the concept of "partial suicide" or "focal suicide."[23] One of the first to address this thesis in any length was Karl Menninger,[24] who also regarded self-mutilation as a means of "fending-off" suicide. Friedman et al.[25] find that self-mutilators usually mount an attack on just one part of the body, and experience a sense of relief following the act. On the other hand, the person attempting suicide attacks his whole body, and contentment, as well as calmness, comes prior to the act. Although this type of classification provides useful organization, the causes of the different actions are much more difficult to formulate.

In the specific case of Vincent we must return to the primary observation that his self-mutilation was not consciously considered. The artist himself was unable to provide an explanation. We are left with a number of plausible but unproven hypotheses, which do no more than accommodate the events. I have presented the more reasonable ones, more or less in order of merit.

The psychoanalysts have had a field day with Vincent's ear-cutting. Although there is nothing inherently wrong with psychological interpretations, such theorists have unfortunately wrapped the artist with thick tissues of imaginative but ill-founded verbiage. The effects of their speculations are still being felt because the press, films, and other popular media reflect their views of Vincent. Vincent's ear-cutting affair certainly must be addressed because it was a bizarre act by any standard. However, it may be one of the least important aspects when properly considered within the total picture of this great artist's unusual life.

References

1. van Gogh-Bonger J. 1978. Memoir of Vincent van Gogh. vol I, XV–LIII, in: *The Complete Letters of Vincent van Gogh*. 2nd ed. Boston: New York Graphic Society.
2. Gauguin P. 1921. *Paul Gauguin's Intimate Journals*. (translated by V W Brooks) New York: Liveright.
3. Favazza AR. 1987. *Bodies under Siege: Self-Mutilation in Culture and Psychiatry*. Baltimore & London: The Johns Hopkins University Press.
4. Sandblom P. 1987. *Creativity and Disease: how Illness affects Literature, Art, and Music*. 4th ed., rev. Philadelphia: G.F. Stickley Company.
5. Gould GM & Pyle WL. 1896. *Anomalies and Curiosities of Medicine*. Philadelphia: W.B. Saunders.
6. Munro WSJH. 1869. On a case of congenital absence of the meatus externus of the right ear. *The Lancet* vol. 2 for 1869: 41.
7. Clarke T. 1981. The van Gogh club, self-mutilation in a Victorian prison: a clinical study. *The Australian Journal of Forensic Sciences* 14: 17–25.
8. Tralbaut ME. 1981. *Vincent van Gogh*. New York: The Alpine Fine Arts Collection Ltd.
9. Doiteau V & Leroy E. 1936. Vincent van Gogh et le drame de l'oreille coupée. *Æsculape* 26th.year (new series), no. 7, pp 169–192.

10. Yasuda K. 1979. Was van Gogh suffering from Ménière's disease? *Otologia Fukuoka* **25**: 1427–1439.
11. Gachet P[L]. 1957. Les médecins de Théodore et de Vincent van Gogh. *Æsculape* **40**: 2–37.
12. Lubin AJ. 1961. Vincent van Gogh's ear. *Psychoanalytic Quarterly* **30**: 351–384.
13. Lubin AJ. 1987. *Stranger on the Earth: A Psychological Biography of Vincent van Gogh.* New York: Henry Holt & Co.
14. Runyan WM. 1981. Why did van Gogh cut off his ear? The problem of alternative explanations in psychobiography. *Journal of Personality and Social Psychology* **40**: 1070–1077.
15. Bonafoux P. 1989. *Van Gogh Self Portraits.* New York: Tabard Press.
16. Zola É. 1983. *La Faute de l'Abbé Mouret* (The Sin of Father Mouret). (translation by S Petrey) Lincoln & London: University of Nebraska Press. (First published in 1875.)
17. Kôdera T. 1990. *Vincent van Gogh: Christianity versus Nature.* Amsterdam & Philadelphia: John Benjamins Publishing Company.
18. Byatt AS. 1990. After the myth, the real. *Times Literary Supplement,* June 29 – July 5, p 683, London: Times Newspapers Ltd.
19. Nagera H. 1967. *Vincent van Gogh: A Psychological Study.* New York: International Universities Press Inc.
20. Graetz HR. 1963. *The Symbolic Language of Vincent van Gogh.* New York: McGraw-Hill Book Company Inc.
21. Schnier J. 1950. The blazing sun. A psychoanalytic approach to van Gogh. *American Imago* **7**: 143–162.
22. Storr A. 1988. *Churchill's Black Dog, Kafka's Mice, and Other Phenomena of the Human Mind.* chapter 11. New York: Grove Press.
23. Walsh BW and Rosen PM. 1988. *Self-Mutilation: Theory, Research, and Treatment.* New York & London: The Guilford Press.
24. Menninger KA. 1935. A psychoanalytic study of the significance of self-mutilations. *Psychoanalytic Quarterly* **4**: 408–466.
25. Friedman M, Glasser M, Laufer E, Laufer M & Wohl M. 1972. Attempted suicide and self-mutilation in adolescence: some observations from a psychoanalytic research project. *International Journal of Psychoanalysis* **53**: 179–183.

Chapter 10
Suicide

Figure 10.1. Vincent van Gogh: *Marguerite Gachet at the Piano*, 1890, Auvers-sur-Oise, oil on canvas, 102.6 × 50 cm (40.4 × 19.7″), Öffentliche Kunstsammlung Basel, Kunstmuseum

To die quietly of old age would be to go there on foot.
Vincent to Theo, letter 506, from Arles, July, 1888.

Late Sunday afternoon, July 27, 1890, Vincent van Gogh shot himself with a borrowed revolver, and returned with difficulty to his room at the Café-Auberge Ravoux, in Auvers-sur-Oise. He died about 1:30 a.m. on Tuesday, July 29. During the interval he was interviewed by several people including the proprietor Ravoux, Drs. Mazery and Gachet, local policemen, and his brother Theo; all of whom discovered very little worth reporting. Ravoux claimed that the revolver[*] was his and had been borrowed by Vincent to "chase crows." The gendarmes were more tiresome than investigative and were dismissed by Ravoux, after Vincent declared that his body was his own to do with as he liked. Anecdotal evidence suggests that Vincent told Dr. Gachet that he would have to do it again if they pulled him through. All of the foregoing supports the notion that Vincent intended to take his own life, but by putting the bullet in his abdomen, rather than his head or heart, he seemingly bungled a rapid demise. Why he did this some distance from his domicile, and why he elected to stagger home rather than use another bullet, are among the many questions that remain unanswered.

Monsieur Ravoux claimed that Vincent needed the revolver to scare off birds in the field. There is no indication that the artist had any experience with firearms and the possibility of an accident should at least be raised if not labored. Furthermore, if he suffered an attack and was hallucinating, he may have directed the revolver at an imaginary bird or other object on his person. Also, in a different vein, there is a rumor[**] in Auvers that young boys shot Vincent accidentally. The story goes that they were reluctant to speak up for fear of being accused of murder and that van Gogh decided to protect them and to be a martyr.

Vincent's death was certainly taken as a suicide by most of the locals. None of the people with any intimacy toward van Gogh spoke otherwise, and the priest in Auvers-sur-Oise even denied use of a cart for the coffin on these grounds. (A brother of the cloth at Méry-sur-Oise was less rigid.) The death

---

[*] The gun was never recovered. Supposedly someone picked it up and was either unaware of its significance or unwilling to surrender it. Within a few years of the centenary of Vincent's death, a revolver showed up in Auvers, and enjoyed momentary candidacy, but it faded from interest for lack of any distinguishing characteristics.

[**] This was first recounted to me in 1988 by Professor John Rewald, who professed no particular belief in its accuracy.

certificate makes no comment on the cause of death and, by today's standards at least, it is curious that neither of the two attending doctors bore official witness. There has been much speculation since on the prevailing circumstances, even though the available data is relatively sparse. Again, the artist's own words provide the most reliable indicators. But before extracting relevant passages from the letters, it is worth summarizing the general and scholarly concepts regarding suicide that have evolved over the ages. I will first endeavor to place Vincent's case in the context of the late nineteenth century and then discuss some general aspects about gifted individuals who commit suicide.

A brief history on attitudes about suicide

St. Augustine (354–430) categorically rejected suicide, *The City of God*, Book I, section 20, as a violation against the Sixth Commandment concerning killing, and because it precluded the possibility of repentance. In the teachings of St. Thomas Aquinas (1225–1274) suicide was depicted as a mortal sin because it usurped God's power over life and death. Most of these Christian injunctions were seemingly a reaction to the classical Roman view of suicide as a neutral, rational response, i.e. that life was rather cheap and should not be continued if the quality deteriorated or, as the Stoics advocated, suicide was a preservation of individual integrity.

John Donne (1573–1631) was one of several English writers who have defended suicide; in *Biathanatos* he stated that man was his own executioner and that this should be taken for granted. (The paradoxical Donne offered the Christian martyrs and even Christ as proof of the legitimacy of suicide, but still avoided judgment about whether it was a sin.) In the same vein, the British physician and author Thomas Browne (1605–1682) called man his own assassin. Shakespeare also allowed that man could connive at his own death as in *Hamlet*, or it might come by some outside force and result in redemptive martyrdom, as in *King Lear*. T. S. Eliot (1888–1965) explores the topic in his play about the martyr Thomas à Becket, *Murder in the Cathedral*.

David Hume (1711–1776) was less metaphorical and in his posthumously published essay "Of Suicide," *Essays and Treatises*, 1777, he tried to decriminalize the act of taking one's own life. The French philosopher Jean-Jacques Rousseau (1712–1778) blamed society for causing generally good and innocent men to take their own lives. This position has been repeatedly rediscovered, and most recently embraced by Antonin Artaud who wrote, "[van Gogh] did not commit suicide in a fit of madness, in dread of not succeeding. On the contrary, he had just succeeded and discovered what he was and who he was, when the collective consciousness of society, to punish him from escaping from its clutches, suicided him."[*]

In his landmark studies on the sociology of suicide, Émile Durkheim (1858–1917) identified three basic types of suicide: *altruistic* (required by society), *egoistic* (resulting from too few ties between subject and community), and *anomic* (caused by sudden shattering of accustomed relationships between the individual and society), which still have utility as operational categories. The psychological flag was carried by Sigmund Freud (1856–1939) who put the locus of suicidal action in man's unconscious mind and emphasized the concept of ambivalence. He not only recognized a duality in the will to live and the will to die, but also recognized destructive feelings that were inwardly directed at an ambivalently viewed love object. Karl Menninger (1893–1990) is credited with delineating chronic suicide (including alcohol addiction) and focal suicide (including self-mutilation).

Edwin Shneidman[1] maintains that the *acute suicidal crisis*, the period of high and dangerous lethality, is an interval of relatively short duration; hours or days, not usually months or years. He also points out that any improvement following a suicidal crisis does not mean that the risk is over; in fact, most suicides occur within about three months following the beginning of "improvement" when the individual has the energy to put morbid thoughts and feelings into effect. (As I note below, three months happens to coincide with a perceived safety period mentioned by Vincent just before he left the asylum at St. Rémy.)

Several authors have emphasized that suicides are typically two-person events. The concept of *the significant other* being able to influence either positively or negatively the potential and the outcome is well documented. Accordingly, suicidal tensions develop between spouse and spouse, parent and child, lover and lover. Generally speaking, suicide imposes the greatest stigma of all modes of death upon its survivors.[1] Vincent's statement is even more to the point: "suicide ... actually turn[s] your friends into murderers" (letter 492).

Part of the mystique that surrounds painters, poets, and writers derives from the idea that because they are sensitive people they are more prone to commit suicide. In point of fact, suicide seems to cross all socioeconomic,

---

[*]   From Antonin Artaud, *Van Gogh, the Man Suicided by Society*, 1947. The English translation is from Sontag S. 1976. *Antonin Artaud, Selected Writings.* New York: Farrar, Straus, and Giroux. Artaud's provocative argument is not without interest, but unfortunately is peppered with inaccuracies such as having Vincent commit suicide while at Dr. Gachet's house. Also, Artaud's negative view of work therapy, and of Dr. Gachet for advocating it, are amusingly paradoxical in the context of his writing this essay while in the asylum at Rodez.

cultural, and geographic boundaries; but with an artist, suicide casts a romantic aura. Much discussion in the literature concerns whether an artist's reputation is elevated following untimely death. It is undeniable that public interest in Vincent van Gogh is stimulated in part by curiosity about his suicide. However, it is equally evident that his position in art history is based on his paintings, the works of a gifted individual with a strong commitment to his identity as an artist.

Indicators of suicide potential in the gifted

In 1921, Lewis Terman[2] initiated a longitudinal study of 1,528 gifted people, who were identified within the public schools of California, at average age 9.7 years, as possessing Stanford-Binet IQ's of 140 or greater. By 1970 there were 28 suicides (20 men and 8 women). A subgroup of five; all male and Caucasian, committed suicide by gunshot, at age 43–58. Shneidman[3] had assistants pull the unseen histories on these five suicides together with ten natural deaths and fifteen living individuals who were matched in terms of age, occupational level, and father's occupational level. In a blinded study, Shneidman successfully rated four of the five suicides at the top of his list of most likely suicide candidates, and the fifth was rated number six. The null hypothesis that there were no discernible prodromal clues to suicide can be discarded with confidence.[3]

The following list of features, which contributed to a "total impression" of a tendency towards suicide, was offered by Shneidman in connection with these five specific cases:

- Early (before age 20) evidences of instability
- Actual or felt rejection by father
- Multiple marriages
- Alcoholism
- An unstable occupational history
- Ups and downs of income and ups and downs of mood
- A crippling physical disability, especially one involving dyspnea
- Disappointment in the use of potential; disparity between aspiration and accomplishment
- Any talk or hint of self-destruction
- A competitive or self-absorbed spouse

Not all of these features occurred in any particular suicidal case. Shneidman made the observation that, "for the five suicidal subjects, for reasons that are not completely clear, it seemed that the relationships with the father were more critical than the relationships with the mother." Vincent van Gogh would surely receive a check mark on this feature, and in fact on most of the categories.

Baechler[4] and others have stressed that suicide should not be considered simply as an act, an illness, or a force, but rather as a behavior, and a solution to a problem. For the suicidal subject his own decision is considered both logical and rational. However, there is a considerable psychiatric content with regard to the mechanism. In one of the most common, the subject feels that he is falling victim to mental illness and finds self-destruction the best possible solution to avoid sinking into madness. I believe that this type of concern contributed heavily to Vincent van Gogh's thinking and drove him to the ultimate action. Furthermore, the overall time-scale, and the relationship of his previous crises to the fatal act, constitute a chronology which demands attention.

A decade devoted to art

There were several indications that Vincent anticipated a relatively short life. Aged 30, writing from The Hague, in mid-1883, he remarked:

> Not only did I begin drawing relatively late in life, but it may also be that I shall not live for so many years. ... my body will keep a certain number of years ... between six and ten for instance [until age 36–40] ... This is a period which I can firmly count on ... If one wears oneself out too much in those years, one doesn't live past forty ... But for the present such calculations are irrelevant; as I said, one can only take plans for a period of between five and ten years into account. I do not intend to spare myself, nor to avoid emotions or difficulties – I don't care much whether I live a longer or shorter time. ... I have walked this earth for thirty years and out of gratitude want to leave some souvenir ... So this is my aim ... all done with one object in mind. ... Guillaume Régamey was somebody who did not leave behind any distinct reputation ... but he was a personality for whom I have great respect. He died at the age of 38, and a six- or seven-year period had been devoted solely to making drawings ... a special example of a certain self-possession and energy clinging to one inspiring idea, of the fact that difficult circumstances showed him the way to accomplish good work in complete serenity. This is the way I regard myself – as having to accomplish something with heart and love in it within a few years, doing this with energy. If I live longer, *tant mieux*, [so much the better] but don't count on it. Something must be done in those few years, this thought dominates all my plans for my work (letter 309).

Vincent's commitment to become an artist can be dated from August 20, 1880 (letter 134). The concept of being an artist for ten years was quoted as an old saying, "You must study for ten years and then produce a few portraits" (letter 542), repeated again "the only thing that comforts me is that people of experience say you must paint ten years for nothing" (letter 607) and again, "Compared with others I still belong to the lucky ones, but think what it must be [like] if one has entered the profession and has to leave it before one has done anything, and there are many like that. Given ten years as necessary to learn the profession and somebody who has struggled through six years and paid for them, and then had to stop, just think how miserable that is, and how many there are like that!" (letter 612). Vincent van Gogh died July 29, 1890, 37 years and 4 months old, just three weeks short of a decade in art. Had he not committed suicide, he could have expected to live another 29 years,[*] for a total of 66 years, and would have lived on until 1919.

Specific timing

In May 1890, just before Vincent's departure from St. Rémy Asylum, he wrote to Theo, "I pointed out to [Dr. Peyron] that such attacks as I have just had have always been followed by three or four months of complete quiet. I want to take advantage of this period to move" (letter 631). We can place the termination of his last crisis fairly accurately at April 29, 1890, because on this day he said, "Until now I have not been able to write you, but being a bit better just now [I put pen to paper]" (letter 629). A safe period of three months would literally terminate on July 29, 1890.

Vincent's expectation of three to four months of peace would fall between July 29 and August 29; or July 22 and August 19 (if one uses four weeks as a measure of a month).[**] The suicidal act was in fact committed on July 27. Thus it is difficult to ignore the hypothesis that Vincent was on that day

---

[*] The estimate of 29.38 years for a 37 year old male in 1890 is actuarial data from Massachusetts, U.S.A., courtesy of Hazel Forbes, Metropolitan Life Insurance Company. Patrick Conan provided an equivalent estimate for France of 28.83, based on an interpolation of data from the Institut National D'Etudes Démographiques. Available information indicates a similar estimate for Holland at that time.

[**] The actual times between major attacks (based on best estimates from comments in the letters) were 38, 148, 116, 21 and 26 days, i.e. showing neither consistency nor a trend. Accordingly, there is little justification for playing with these numbers, but the mean happens to be 70 days (standard deviation 58) and the next attack would be predicted to fall between May 11 and September 4, centered about July 8.

either in the throes of another crisis or responding to prodromal signs of an impending crisis. It is worth noting that Johanna van Gogh-Bonger[5] was of the opinion that "fear of an impending attack or the attack itself drove him [Vincent] to death."

It is of interest to ask whether this time frame was perceived by Theo and others. In his second letter from Auvers-sur-Oise, Vincent mentioned that "I can do nothing about my disease. I am suffering a little just now" (letter 636), and in letter 648 (which Hulsker[6] feels should be placed between 636 and 637), while discussing the consequences of a possible cessation of painting, he ends with "the prospect grows darker, I see no happy future at all." Theo seemed to be aware of possible relapses because Johanna van Gogh-Bonger[5] recalled that on June 20, 1890, Theo wrote to her in Holland, "If only Vincent is not getting melancholy and a new crisis is to be expected, everything [else] has gone so well lately." Perhaps the closest Vincent came to issuing another warning was on June 30, to both Theo and Jo: "I myself am also trying to do as well as I can, but I will not conceal from you that I hardly dare count on always being in good health. And if my disease returns, you would forgive me" (letter 646). In retrospect we can see several warnings to Theo. But he did not respond to them, in part at least because he was distracted by his own problems both at home and at work.

Vincent's sentiments on death and illness

One of the most beautiful passages in Vincent's letters was written in August 1888, from Arles:

> Perhaps death is not the hardest thing in a painter's life. For my own part I declare that I know nothing whatever about it, but looking at the stars always makes me dream, as simply as I dream over the black dots representing towns and villages on a map. Why, I ask myself, shouldn't the shining dots of the sky be as accessible as the black dots on the map of France? Just as we take the train to get to Tarascon or Rouen we take death to reach a star. One thing undoubtedly true in this reasoning is that we cannot get to a star while we are alive any more than we can take a train when we are dead. So to me it seems possible that cholera, gravel, tuberculosis and cancer are the celestial means of locomotion, just as steamboats, buses and railways are the terrestrial means. To die of old age would be to go there on foot (letter 506).

It demonstrates not only an ability with the written word, but also a very personal declaration about life and death. There were also numerous references by Vincent of wholesome adjustments to his illness – for example, "The difference between happiness and unhappiness! Both are necessary and useful as well as death or disappearance ... it is so relative – and life is the same. Even faced with an illness that breaks me up and frightens me, that belief is unshaken" (letter 607).

On the other hand it is unsettling to read in van Gogh's letters the recurring theme that sickness, or something less than full health, is compatible with creative work, for example, "one must not think that people whose health is impaired, wholly or partly, are not good for painting ... nervous people are more sensitive and refined" (letter 449), or "it isn't necessary to be a Hercules to stick assiduously to the easel" (letter 604). Vincent even suggested that sickness encourages creativity: "Now if I recover I must begin again, and I shall not again reach the heights to which sickness partially led me" (letter 570). This was an unfortunate philosophy, relating the fevers of illness to activities on a higher plane, an idea that seems to have been quite prevalent in the eighteenth and nineteenth centuries.

Vincent's thoughts of suicide

Early references to suicide indicated feelings of repugnance, inherent dishonesty, and unacceptability (letters 154, 212, 268, 462). On May 29, 1888, he wrote to Theo from Arles: "Remember that I would far rather give up painting than see you killing yourself to make money. ... can't you see that ... living for other people is a mistake if it involves suicide, for in that case you actually turn your friends into murderers" (letter 492).

Vincent's first crisis at Christmas 1888, which included the ear-cutting incident, might be considered a "focal suicide" or a means of fending off suicide in Menninger's terminology.[7] On April 10, 1889, he alluded to stronger notions when he wrote euphemistically to sister Wil, "I am unable to describe exactly what is the matter with me ... Every day I take the remedy which the incomparable Dickens[*] prescribes against suicide. It consists of a glass of wine, a piece of bread with cheese, and a pipe of tobacco" (letter W11). In March 1889, when Vincent was locked up in a cell in response to a public petition, he declared that "I would rather have died than have caused and suffered such trouble" (letter 579).

However, perhaps the first rationalization was expressed on April 30, 1889, "If I were without your [Theo's] friendship, they would remorselessly

---

[*] Charles Dickens (1812–1870), English novelist and, incidentally, one of the world's greatest hypochondriacs.

drive me to suicide and, however cowardly I am, I should end by doing it. There ... is [a] juncture where it is permissible for us to protest against society and defend ourselves" (letter 588). There were no direct statements on self-destruction from St. Rémy. However, Dr. Peyron wrote to Theo, around August 1889, "His thoughts of suicide have disappeared, only disturbing dreams remain" (letter 602a). The concept that only courage was lacking surfaced again on September 10, 1889: "Life passes like this, time does not return, but I am dead set on my work, for just this very reason that I know the opportunities of working do not return. Especially in my case in which a more violent attack may forever destroy my power to paint. ... I am now trying to recover like a man who meant to commit suicide and finding the water too cold tries to regain the bank" (letter 605). In the same letter he wondered whether he should have been more forceful in Arles: "I ought ... to have defended my studio ... Others in my place would have used a revolver ... as it is I've been cowardly and drunk." I am convinced that the underlying disease was still the most important influence, "during the attacks it is terrible – and then I lose consciousness of everything. But that spurs me on to work and to seriousness, like a miner who is always in danger makes haste in what he does" (letter 610). The driving force and the sense of impending doom are palpable.

Last letters

Vincent's mother and his sister Wil received their final letter from Auvers-sur-Oise within two weeks of his death; it is inscribed in the mother's handwriting, "Very last letter from Auvers." He depicted what now appears as a calm before a storm, "For the present I am feeling much calmer than last year, and really the restlessness in my head has greatly quieted down. ... I myself am quite absorbed in the immense plain with wheatfields against the hills, ... I am in a mood of almost too much calmness, in the mood to paint this" (letter 650).

Johanna van Gogh-Bonger annotated letter T41 (from Paris, July 14) as Theo's last communication sent to Vincent; it specifically mentions enclosing a 50-franc stipend. However, there is good reason to suppose that Theo wrote again, probably about July 22, after the anticipated eight days[*] for his return from Holland (see close of letter T41). This would be in keeping with Vincent thanking him for a letter and a 50-franc note received on July

---

[*]   Actually, Theo returned at least two days earlier, and wrote from Paris to his wife in Holland on July 20.

268

23 (letter 651). (In those days, mail between Paris and Auvers was delivered on the same day or the next day.) It is unfortunate that this letter from Theo has been either suppressed or lost. There were indications of strained relationships among Theo, Johanna, her brother and sister-in-law (the Bongers who lived in the same apartment building in Paris), at that time and perhaps the letter contained unfortunate or embarrassing, but not necessarily sinister, comments which the family felt should not be published. The simplest explanation is that it was lost, which would be in keeping with the fate of the great majority of Theo's letters to Vincent. In the same vein, sentences three through seven of Vincent's letter 651 have been omitted in *The Complete Letters*[5] compared with the translation of the original offered by Hulsker,[6] which I assume now completes that letter.

Letter 652 was found on Vincent after he shot himself. It was neither an unfinished communication nor a suicide note, as has been suggested in several popular press articles. It seems most likely to have been a first draft[6] of letter 651. According to Theo's later inscription it was "found on him on July 29," supposedly after Vincent died. Had Vincent inadvertently shoved it in his pocket a week earlier? Did he have a purpose in keeping it? Is there any indication of Vincent drafting letters before this one? These questions spring to mind but remain unanswered. Here are the two documents, side-by-side, with artificial spacing to highlight similarities and differences.

letter 651, July 23, 1890, supposedly the last letter received from Vincent by Theo.

letter 652, putative draft for letter 651, found on Vincent by Theo, July 29, 1890.

My dear brother,

My dear brother,

Thanks for your letter of today and the 50-fr. note it contained.

Thanks for your kind letter and for the 50-fr. note it contained.

Perhaps I'd rather write you about a lot of things, but to begin with, the desire to do so has completely left me, and then I feel it useless.

There are many things I should like to write you about, but I feel it useless.

I hope that you will have found those worthy gentlemen well disposed toward you.

I hope you have found those worthy gentlemen favorably disposed toward you.

As to the peace in your household, I am equally convinced of the possibility of conserving it, as of the storms that threaten it. I prefer not to forget the little French I know and I certainly can't see the usefulness of stressing the wrong or the right on both sides in eventual discussions. Only it wouldn't interest me. Here things go fast – Dries, you and me, are we not a little more convinced, don't we feel it a little better than the ladies? Good for them – but after all, to discuss things quietly, we don't even count on it anymore.

Your reassuring me as to the peacefulness of your household was hardly worth the trouble I think, having seen the wheal and woe of it for myself. And I quite agree with you that rearing a boy on the fourth floor is a hell of a job for you as well as for Jo.

As far as I'm concerned, I apply myself to my canvases with all my mind, I am trying to do as well as certain painters whom I have greatly loved and admired.

Now I'm back, what I think is that the painters themselves are fighting more and more with their backs to the wall.

Very well ... but isn't the moment for trying to make them understand the usefulness of a union already gone?
On the other hand a union if it should take shape would founder if the rest should have to founder. Then perhaps you would say that some of the dealers might combine on behalf of the impressionists but that would be very short-lived. Altogether I think that personal initiative remains powerless, and having had experience of it should we start again?

I noticed with pleasure that the Gauguin from Brittany which I saw was very beautiful and I think that the others he has done there must be so too.
Perhaps you will look at this sketch of Daubigny's garden. It is one of my most purposeful canvases. [Further descriptions of a sketch]
[Orders for paints on behalf of Hirschig]
[Directions to paint dealer]
I have reduced my own order [for paints] to the barest minimum.
[Comments on Hirschig and Konig]

Since the thing that matters the most is going well why should I say more about things of less importance?
My word, before we have a chance to talk business more collectedly, we shall probably have a long way to go.
The other painters, whatever they think, instinctively keep themselves at a distance from discussions about the actual trade.
Well, the truth is we can only make our pictures speak.

But yet, my dear brother, there is this that I have always told you and I repeat it once more with all the earnestness that can be expressed by the effort of a mind diligently fixed on trying to do as well as possible – I tell you again that I shall always consider you to be something more than a simple dealer in Corots, that through my mediation you have your part in the actual production of some canvases which will retain their calm even in a catastrophe. For this is what we have got to, and this is all, or at least the main thing, that I can have to tell you at a moment of comparative crisis. At a moment when things are very strained between dealers in pictures of dead artists and living artists.
Well, my own work, I am risking my life for it and my reason has half foundered because of it – that's all right – but you are not among the dealers in men as far as I know, and you can still choose your side I think, acting with humanity, but que veux-tu [what can we do]?

[conclusion of letter 652]

Good-by now, and good luck in business, etc.,
remember me to Jo and handshakes in
thought.

Ever yours, Vincent.

Both documents acknowledge receipt of the stipend. Both allude to many unidentified items which Vincent feels are useless to discuss. For Theo, this was surely a hollow, annoying statement. It takes on more significance for us because it resisted editing.

The almost identical comments about the "worthy gentlemen" refer to the management of the Goupil Gallery. Hulsker[6] has pointed out that Theo had declared to his mother and Wil on July 22 that he would remain with the company, but for reasons unknown Vincent was either uninformed or still left in doubt about Theo's decision. Then, Theo's domestic problems are actually expanded in the mailed version.

I have assumed that the things that are "going well" in the draft can be identified with "my canvases" in the letter. The other reasonable possibility is that Vincent was referring in the draft to the recovered health of his nephew. Both documents go on to decry the withdrawal of struggling artists from the commercial fray. In the letter, we see the re-emergence of the "society of artists" concept, which Vincent started in Holland and took with him to Arles. It is somewhat surprising at this date, the more so because it is not mentioned in the draft. This must surely be related to recent discussions with Theo on his prospects for an independent dealership and his brother's already established relationships with the Impressionists. Perhaps this was prompted by something in the lost letter from Theo?

The draft contains a solid endorsement of Theo as both a provider for Vincent and a creator of art, albeit once removed. This theme was deleted in the final copy, perhaps because Vincent felt that it was too self-serving. The reference in the draft to Theo being "a simple dealer in Corot's" springs in part from Theo's letter of July 14 in which he enthusiastically relates hopes of selling paintings by Corot and Diaz after his swing through Leiden to accompany his wife and son to his mother's home. Vincent's choice of these two artists, which he much admired, and his next reference to the conflict between promoting dead artists compared with living artists such as himself, must have been particularly onerous when discovered by Theo. In other words, there was good reason for Vincent's election to omit this from his mailed letter. Instead he seems to have substituted art discussions of a type quite typical of past correspondence, together with some orders for pigments for himself and his colleague. It is worth noting that his minimum order of materials contrasts in mood to recently described successful canvases.

It is difficult and perhaps dangerous to read more into these documents, especially with regard to the draft version. The gloomy aspect in the latter about risking life for work obviously cannot be ignored, but does represent a progression of past sentiments. Notwithstanding the obvious problems facing the brothers, neither delivered letter nor discovered draft actually connotes impending doom or a cry for help.

Contributing factors to Vincent's suicide

In chapter 2, I alluded to the unsettled household of Theo van Gogh in Paris during Vincent's last two months in nearby Auvers-sur-Oise. At the end, partly due to a failure in communications, Vincent was not aware of Theo's most recent decision to keep his present employment and to delay branching out on his own – something that Vincent had continually urged in the past but found inopportune in mid-1890. Accordingly, the artist undoubtedly felt somewhat threatened in terms of a lessening of financial support from his brother, and this contributed to his bleak outlook. However, I am not inclined to take financial uncertainty as a primary reason for suicide because we must offset this potential loss with at least a beginning of sales for Vincent's work and indications of sustenance support from Dr. Gachet and others. Moreover, Vincent's expressed fears of being a financial burden to Theo were apparently effectively relieved by Johanna (letter 649) and arguments to the contrary become somewhat circular. Vincent had gotten by for ten years. He was not completely incapable of earning a living, and had even suggested, a little over a year before, that he might do a stint in the French Foreign Legion.

Claims that Vincent was aware of Theo's failing health in 1890, and wanted to decrease his brother's burden, surfaced in newspaper speculations one hundred years later. The artist presented no evidence of this in letters. Even if Vincent had some unexpressed insight on an early demise for his brother, the consequences of the loss of both men to the widow Johanna and his namesake nephew would surely not have escaped him. In chapters 5 and 6, I presented the case for Theo also suffering from acute intermittent porphyria, whether or not his condition was exacerbated by syphilis. The suggestion that Vincent's suicide was in sympathy with Theo's illness or other problems is not supported by extant facts.

A relationship to acute intermittent porphyria

Gastrointestinal pain during crises is a hallmark of acute intermittent porphyria. Vincent certainly had a history of such complaints, and this has been addressed in chapters 3 and 5. If he was suffering from another crisis on the afternoon of July 27, it is possible that abdominal pain influenced the direction of his revolver. Especially if he were hallucinating, we now have a rather simple but novel explanation for the target organ. There is another datum which lends a little credence to this idea. Unfortunately, it appeared many years after the fact, but the Dutch newspaper article by Dr. A. Bredius, referenced by Tralbaut,[8] concerns the recollections of Anton Hirschig, the neophyte artist, colleague to Vincent, and Dr. Gachet's messenger to Theo. According to Hirschig's memory of Vincent, immediately after the artist's return to the Café-Auberge: "I can see him in his little bed in his little attic, in the grip of terrible pain. 'I couldn't stick it any longer so I shot myself,' he said. *'But will nobody cut my belly open for me?'* (my italics). It was swelteringly hot up there under the roof." Later in the evening the pain seemed to disappear and Vincent supposedly puffed away at the pipe that Dr. Gachet had lit for him.

Other hypotheses

A romantic hypothesis revolves around Vincent's possible interest in Marguerite Gachet, the doctor's only daughter and subject of a van Gogh painting, *Marguerite Gachet at the Piano.*[*] Dr. Gachet is thus cast as the objecting father who could tolerate Vincent as a great artist but not as a son-in-law. Anecdotal local evidence, unfortunately rather removed from the time of the suicide, suggests that Mademoiselle Gachet withdrew from the world because Vincent was gone. However, others have suggested that both daughter and son were preoccupied with preserving a monument to their father.

Tralbaut,[8] Lubin,[10] Nagera,[11] and Schnier[12] have assembled a number of working hypotheses and should be consulted directly for views with psychological spin. I will mention a few of these, but only briefly. One suggestion is that Vincent wanted "to go out on the crest" of a creative wave. For the adherents of this hypothesis, Vincent's fears of success or of decreasing skills were close to the surface, but not visible to most of us. Many

---

*   Roger Golbéry,[9] a distant relative by marriage, reproduced a photograph of Marguerite Gachet at the same piano in 1947.

commentators continue to invoke Vincent's unresolved problems with his deceased father; this may have contributed to long term behavior but was not evident at the end. The specter of Vincent's stillborn older brother, and his gravestone at Zundert, gets a lot of mileage as a sinister background influence. However, there is no mention of either in Vincent's letters. By dying in his mid-thirties, some have even suggested that Vincent was following the example of Jesus Christ. And so it goes.

The suicide of brother Cornelis

Vincent's youngest sibling, Cornelis van Gogh, was born in Zundert on May 17, 1867, and died in South Africa, April 12, 1900. The circumstances of his death are not completely clear, but the possibility of suicide has been entertained. The information is based upon comments within *The Complete Letters of Vincent van Gogh*,[5] and a paper by Ploeger,[13] written in Afrikaans. Although Colonel-Doctor Ploeger admits to significant gaps in the history, his findings merit restating in some detail because the original article is relatively inaccessible,[*] and has been given only cursory coverage by other commentators.

Cor was born to a 47-year-old mother and a father of 45. The eldest sister, Anna, married when he was eleven. He was the youngest of six children, and somewhat pampered in his adolescence. His eldest brother Vincent was already 14 when Cor was brought into the world, and he seems to have taken only a casual interest in his youngest brother as judged from infrequent references in letters. Vincent did send Cor two fine-art prints in 1877, one for Cor's tenth birthday (letter 95), and the other at Christmastide (letter 115), and he also recounted enjoyable moments with the youngster while sledding in the snow (letter 116a). Eleven year old Cor had the charming and unique distinction of drawing on the same sheet of paper with Vincent. On a little illustrated map of the Etten district, towards the end of July 1878, he wrote for Theo, "Vincent and I did this in the pine wood – Cor – I must [now] go to bed. Good night" (letter 123).

In the first half of 1882, Vincent expressed passing interest in Cor's education (letter 184). He was sent to a boarding school in Helmond, performed well, and graduated in mid-1884 at age 17. He subsequently worked in the same town, a little northeast of Eindhoven, as an apprentice fitter and turner. About that time Vincent started wondering whether Cor

---

[*] I am indebted to Mr. Jan Hellings for providing a copy of this paper, and to Dr. Jessica Hellings for the English translation.

should join the Goupil Company as an art dealer, notwithstanding his own unhappy experience. His closing comment – "At present I think him [Cor] a nice boy, but it is certainly time for him to do something practical" (letter 380) – must have impressed Theo as coming from a dubious role model.

From Arles, in February 1888, Vincent wrote, "I am glad that our brother Cor has grown bigger and stronger than the rest of us" (letter 518). We have no data to substantiate this evaluation or any further records on Cor's health. Sometime in 1889, Cor contracted with Nederlandsch Zuid-Afrikaansche Spoorweg Maatschappij (The Dutch South-African Railway Company), headquartered in Amsterdam, to work as an engineer in the Transvaal. Vincent wrote a philosophical letter to his mother comparing Cor's anticipated trip to Africa with tales by Paul Gauguin about Central and South America and comments on Australia (supposedly by John Russell); places he felt should be much in common (letter 598).

Cor's departure by steamship from Southampton, England, occurred in the second half of August 1889; he had visited Theo and Johanna in Paris on the way (letter T15). Mother van Gogh received a letter from Cor in September, en route to South Africa (letter T16). He arrived in Johannesburg (letter T19) but soon took up residence in Pretoria where he worked in the railroad central workshop.[13] Vincent was optimistic: "thank you for your last letter and the news of Cor's good voyage. I believe that he will work there with zest and thus have some pleasure in his life" (letter 612). But early reports, according to Theo, were not particularly happy: "Cor writes often from the Transvaal. Life over there can hardly be very amusing. There are no plants or flowers ... One day is absolutely like another, which is why he says he detests Sundays and other times of leisure" (letter T22). There is no indication that Vincent wrote directly to Cor in South Africa but he kept in touch via other members of the family.

In February 1898, Cor van Gogh married a 20-year-old girl, Anna Fuchs, born in Germany but resident in Pretoria for 16 years. The marriage was unsuccessful; within eight months she left him, apparently taking most of the furniture. Her departure prompted Cor to write a will in October 1898, naming his mother as sole beneficiary. In the event his mother should not survive him, he instructed that the estate should go to his sister Wil. His brother-in-law, Joan Marinus van Houten (married to sister Anna), was to be executor. After Cor's death during the Boer War, the Supreme Court in Pretoria thwarted his intent by appointing his estranged wife as executrix. Much correspondence then flowed among van Houten, the Transvaal court, and the railway company officials in Amsterdam as well as Pretoria. Anna van Gogh-Fuchs continued to live in Pretoria after the war, and contested her husband's will until September 1905. Ploeger should be congratulated for following this paper chase with diligence, but the end was anticlimactic;

mother van Gogh received 16 pounds and 18 shillings from the Railway Company (Cor's back pay), van Houten wrote off Cor's outstanding debt to him of more than 80 pounds, and Anna van Gogh-Fuchs held on to the furniture.

Cor's remaining two years after separation from his wife were much influenced by the Boer War (1899–1902). According to Ploeger, many Dutch expatriates in the Transvaal volunteered to fight against the British as early as September 1899. Cornelis seems to have been drawn into the conflict a little later, in response to some of the larger commercial establishments in Johannesburg and Pretoria running full page advertisements against the Boers. This elicited a series of local petitions, including one signed by Vincent's brother. He eventually enlisted in the Volunteer Corps of General J. S. F. Blignaut, in early 1900. He died a few months later in Brandfort, a small town south of Pretoria, in the Orange Free State.

The details have not surfaced, and Ploeger[13] is not sure whether Cor died in the township or in the surrounding district. The death certificate issued on November 22, 1900, by E. M. Ellenberger, of the information bureau of the Transvaal Red Cross, states that Cornelis van Gogh died by accident during an illness with fever on April 12, 1900. An officer under General Blignaut wrote a covering letter to accompany Cor's personal belongings; he also referred to April 12 as the death date. In another document, a field officer (Melt Marais of Pretoria) noted Cor's date and place of birth accurately, that he had been married to Anna Fuchs, left a will, and had died at age 32.75 years, on April 14 (sic). Ploeger[13] pursued the available information on Cor's demise to P.L.A. Goldman's Post-War Register of Military Deaths, which he personally regards as well researched. Therein he found another cryptic entry under deaths as a result of accident: "Van Gogh / Cornelis / 32 / Pretoria Z.A.S.M. [Zuid-Afrikaansche Spoorweg Maatschappij] / 12 April [19]00 / suicide during fever / according R.C. [Red Cross]." This is the only formal evidence for Cor's death by suicide. Perhaps the most interesting datum is the mention of fever in two of these accounts; it is reminiscent of Johanna van Gogh-Bonger's expression,"a state of violent excitement, an attack of high fever," which she used in describing Vincent's ear-cutting incident.

A recent letter from C. A. Venter, town clerk of Brandfort, to my colleague Jan Hellings, confirmed most of the above information but compounds the problem by giving the date of death as April 14. I wonder if this is the burial date? The clerk added that there is a strong local opinion that Cornelis van Gogh is buried in one of the few unmarked graves in the town. Proper records of burials were not maintained before August 1903. An unofficial, illustrated poster item from the same office states categorically, "[Cor] was taken prisoner by the English forces, became ill, and committed suicide in

Brandfort Hospital." We are left with the intriguing possibility that Cor suffered from a medical crisis that led to his taking his own life, and that he may have suffered from the same underlying illness as his brothers. A commemorative tablet in a Dutch Reformed Church in Pretoria is dedicated to expatriates who fell during the war of 1899–1902. Half way down the first column is the name C. V. van Gogh.

Summary

There can be no doubt that Vincent van Gogh experienced terrifying medical crises. He had thoughts and made communications concerning suicide. The journey north from the asylum at St. Rémy to the environs of Paris was undertaken in a self-declared window of safety estimated at three to four months. His death occurred approximately three months after recovery from the last crisis in St. Rémy. The available evidence supports the hypothesis that Vincent either suffered another attack on July 27, 1890 and accidently shot himself or, alternatively, that he was not prepared to go through another crisis which either started at that time or he felt was impending. Problems associated with Theo and Theo's family, or arising from Theo's future employment and Vincent's prospects for continuing financial support, were of secondary importance.

References

1. Schneidman ES. 1980. Suicide. chapter 44, in: *Death: Current Perspectives*. 2nd ed. ES Schneidman (ed.) Palo Alto, California: Mayfield Publishing Company.
2. Terman LM. 1925. *Genetic Studies of Genius: I. Mental and Physical Traits of a Thousand Gifted Children*. Palo Alto, California: Stanford University Press.
3. Shneidman ES. 1980. Suicide among the gifted. chapter 46, in: *Death: Current Perspectives*. 2nd ed. ES Schneidman (ed.) Palo Alto, California: Mayfield Publishing Company.
4. Baechler J. 1979. *Suicides*. (translated by B. Cooper) New York: Basic Books Inc.
5. *The Complete Letters of Vincent van Gogh*. 1978. 2nd ed. Boston: New York Graphic Society.
6. Hulsker J. 1990. *Vincent and Theo van Gogh: A Dual Biography*. Ann Arbor: Fuller Publications.
7. Menninger KA. 1935. A psychoanalytic study of the significance of self-mutilations. *Psychoanalytic Quarterly* 4: 408–466.
8. Tralbaut ME. 1981. *Vincent van Gogh*. New York: The Alpine Fine Arts Collection Ltd.
9. Golbéry R. 1990. *Mon Oncle, Paul Gachet*. Paris: Editions du Valhermeil.
10. Lubin AJ. 1987. *Stranger on the Earth: A Psychological Biography of Vincent van Gogh*. New York: Henry Holt & Co.

11. Nagera H. 1967. *Vincent van Gogh: A Psychological Study.* New York: International Universities Press.
12. Schnier J. 1950. The blazing sun. A psychoanalytic approach to van Gogh. *American Imago* 7: 143–162.
13. Ploeger J. 1981. Cornelis Vincent van Gogh in Transvaal. *Lantern: Journal of Knowledge, Art, and Culture* December, 1981: 51–59.

Chapter 11
Creativity and the van Gogh Legacy

Figure 11.1. Vincent van Gogh: *The Starry Night*, 1889, St. Rémy, oil on canvas, 73.7 × 92.1 cm (29.0 × 36.3″), The Museum of Modern Art, New York. Acquired through the Lillie P. Bliss Bequest

*I can very well do without God both in my life
and in my painting, but I cannot, ill as I am,
do without something which is greater than I,
which is my life – the power to create.*
Vincent to Theo, letter 531, from Arles,
September, 1888.

Vincent was buried on July 30, 1890, and a devastated Theo van Gogh returned to Paris. In the days that followed he took stock of all the paintings and drawings by Vincent, several canvases by other artists which they had collected together, and a pile of letters from his brother. Although Vincent did not leave a will, there was no doubt that the art work should belong to Theo, and Theo alone. We need not enumerate all of the letters in which Vincent stated that his stipends from Theo were to be offset by the production of pictures. Even the note that was found at the end, "through my mediation you [Theo] have your part in the actual production of some canvases" (letter 652) repeats that sentiment. Theo had already shared several pictures with Dr. Gachet and other mourners at the funeral, and there was no hint of selfishness. Nevertheless, Theo did take the time to clear the inheritance with his mother, brother Cor, and sisters, in a manner that reflected business acumen. The rest of the family, and most of their contemporaries, were simply not that interested in the canvases.

Theo's immediate mission was to promote Vincent's art. He even contemplated publishing some of the letters. But illness thwarted his efforts. Upon his death the following year, the responsibility for both collections fell to his widow Johanna van Gogh-Bonger. She set about organizing the letters and eventually completed the translation of 526 of them into English. She was assisted in the English phrasing and idiom by Helen Apel Johnson.[1] The remaining items were completed by the artist's nephew.[2] Notwithstanding the magnitude of this task, they claimed[3] some intentional delay because Johanna van Gogh-Bonger felt that "it would have been unfair to the dead artist to arouse interest in his person before the work, to which he had sacrificed his life, was recognized and appreciated as it deserved to be." Rewald[3] has described the evolution of the different printed versions of the letters with respect to number and language, and the culmination of the project with the publication of the three volume set in English (1958).

Johanna van Gogh-Bonger received very little assistance with the art collection. Père Tanguy continued to provide something of a Mecca for impressionist and postimpressionist contemplation, but he was hardly a promoter or salesman of art, and in any case died in 1894. The liquidation sale after Tanguy's death realized only 130 francs for two van Gogh paintings.[3] Other Parisian dealers were reluctant to embrace unfashionable artists. Johanna's brother Andries advised her to get rid of the van Gogh collection forthwith. The unfortunate position she found herself in was far from unique. With the notable exception of Claude Monet, few of the Impressionists had good sales through the 1880's, and those without another source of income lived no better than hand-to-mouth. The so-called Impressionist exhibitions had made nobody rich.

The Société Anonyme des Artistes, Peintres, Sculpteurs, Graveurs, etc., as the group of Impressionists initially called itself, made an attempt to get together as early as 1867, but it was not until the Spring of 1874 that the first of their organized exhibitions took place. They recurred at irregular intervals. The disparate ideas and ambitions of its members could be accommodated only within the loosest of confederations: the talented group continually split into factions. The only reasonable opportunity for Vincent to participate would have been the eighth and last show in 1886, but he was

Table 8. Exhibitions of the Impressionists*

	1874	1876	1877	1879	1880	1881	1882	1886
total number of artists	30	19	18	14	18	13	9	17
works/exhibition	167	252	241	225	232	170	203	247
works/artist [high]	11	36	31	38	37	34	36	27
works/artist [low]	2	8	2	2	2	1	9	7
selected artists								
Mary Cassatt				√	√	√		√
Paul Cézanne	√		√					
Edgar Degas	√	√	√	√	√	√		√
Paul Gaugin				√	√	√	√	√
Armand Guillaumin	√		√		√	√	√	√
Claude Monet	√	√	√	√			√	
Berthe Morisot	√	√	√		√	√	√	√
Camille Pissarro	√	√	√	√	√	√	√	√
Lucien Pissaro								√
Auguste Renoir	√	√	√				√	
Georges Seurat								√
Paul Signac								√
Alfred Sisley	√	√	√				√	
Victor Vignon					√	√	√	√

* Data extracted from Moffett CS. 1986. The New Painting: Impressionism 1874–1886. San Francisco: The Fine Arts Museums of San Francisco.
** Participation is indicated with a check mark.

neither ready nor invited. In Table 8, I have indicated some of the individual participants in the exhibitions of the Impressionists; my selection is based in large part upon references to these artists elsewhere in the narrative. Other items demonstrate fluxes in the numbers of exhibitors and works of art.

Belated recognition

In chapter 1, I mentioned the few sales and the numerous exchanges that Vincent made during his career. Most commentators feel that the sale of *The Red Vineyard* in 1890 was the most significant in his lifetime. Vincent had been invited, through Theo, to show at a group called Les Vingt, [The Twenty] founded in Brussels in 1883* under the guiding hand of Octave Maus, an attorney and amateur artist. Maus had been tutored in music by his cousin Anna Boch, sister to Vincent's friend Eugène. She was an accomplished pianist and painter and is also remembered as the purchaser of *The Red Vineyard*. This sale was certainly pleasing to both Theo and Vincent. Another satisfying experience was participation in the exhibitions of the Independents. I have indicated in Table 9 the years of these exhibitions and some of the participants; once again the examples have been chosen primarily by dint of reference in earlier chapters.

Vincent's part of the show in 1890 was particularly well received. While the artist was suffering his last medical crisis in St. Rémy, Theo was able to send good news from Paris. "Diaz [sic, should read Duez]** stopped me [Theo] in the street and said, 'Give your brother my compliments and tell him his pictures are highly remarkable.' Monet said that your [Vincent's] pictures were the best of all in the exhibition ... [Charles-Emmanuel] Serret [1824–1900] ... was enraptured" (letter T32).

Albert Aurier had emerged as a champion for Vincent with his article "Les Isolés" in *Mercure de France*, January 1890. Vincent's initial response was self-effacing (letter 625, February 1, 1890), but he liked Aurier's article sufficiently to ask Theo to send copies to Alexander Reid, H.G. Tersteeg, and their Uncle C. M. (letter 626). Accordingly, it was not surprising that

---

* In November 1893 the membership of *Les XX* voted to end their exhibitions, with the understanding that their avant garde task was accomplished. Shortly thereafter Octave Maus founded another society, *La Libre Esthétique,* which was devoted to writers, poets, and art criticism but not artists per se.

** Ernst-Ange Duez (1843–1896) showed with the Independents. Vincent much admired Narcisse Diaz de la Peña (b. 1807 or '8), who was a member of the Barbizon school and referenced elsewhere in the van Gogh correspondence, but Diaz died in 1876.

Table 9. Some Participants* in the Exhibitions of the Independents.

	1884	1886	1887	1888	1889	1890	1891	1892	1893
Émile Bernard							√	√	
Dr. Paul Gachet**							√	√	
Vincent van Gogh***				√	√	√	†		
Armand Guillaumin	√						√	√	
Lucien Pissaro		√	√	√	√	√	√	√	√
Georges Seurat	√	√	√	√	√	√	√	†	
Paul Signac	√	√	√	√	√	√	√	√	√
Henri Toulouse-Lautrec					√	√	√	√	√

* Thirty one artists took part in one or more exhibitions. Signac was among three men to show work all nine events.[4]
** Dr. Gachet had 10 works (the maximum allowed) in both the 1891 and 1892 shows of the Society of Independents.
*** Vincent van Gogh had 3 paintings and some drawings in the 1888 exhibition, two paintings in 1889, and 10 paintings in both 1890 and 1891.
√ = participation; † = retrospective exhibition.

Theo invited Aurier in August 1890 to "write a biography [on Vincent] for which I [Theo] could furnish all the material, which would be all the more veracious because I have in my possession a nearly consecutive correspondence with him [Vincent] from 1873 onward, as well as several interesting documents" (letter T55). Aurier was not averse to the assignment but died from typhoid fever at the age of 27 in 1892. In that same year Theo's widow exhibited 100 van Gogh oils and drawings at the Panorama Gallery in Amsterdam, and Émile Bernard organized a van Gogh memorial exhibition at Le Barc de Boutteville Gallery, with only modest results.

The first significant exhibition of Vincent's work appeared in the Bernheim-Jeune Gallery, in Paris, 1901. It impressed many practicing painters of the time, not least of which Derain, Vlaminck, and Picasso. According to Rewald,[3] the first public gallery to acquire a van Gogh was the Folkwang Museum at Essen, *A Corner of the Asylum Garden;* Figure 5.6, this is the picture about which Vincent referred to the red and black or manic and depressive sensations (letter B21). Thanks to the early promotion by Paul and Bruno Cassirer in Berlin, many paintings by Vincent were sold into private hands as well as museums. The favorable reception of van Gogh paintings in Germany reached a climacteric* in the Cologne exhibition of 1912, which

---

* The provenance[5] of the first of the two portraits of Dr. Gachet provides an interesting but pathetic contrast. The picture found its way from Johanna van Gogh-Bonger through various commercial galleries in France and Germany and

included 108 paintings and 16 drawings.[3] The earliest shows in England were due to Roger Fry (London, 1910), and Oliver Brown (London 1923, 1926). The Museum of Modern Art in New York showed van Gogh in 1929 and 1935.

Johanna van Gogh-Bonger sold Vincent's paintings partly out of financial necessity but also with a wholesome desire to distribute his work beyond France and Holland. Without this the exponential increase in the fame and image of Vincent van Gogh would have been severely inhibited, and thus Vincent's sister-in-law initiated the international recognition that shows no sign of waning. She also deserves worldwide acclaim for her efforts in organizing and translating the letters, projects which were brought to fruition in New York City. After her death in 1925, her son and the artist's nephew, V. W. van Gogh, who had already assumed a less generous and more nationalistic posture, was reluctant to place any more canvases or drawings on the open market. He eventually worked out an agreement with the Dutch government involving a substantial family settlement, a new state museum to house the donated works, and influence on the administrative van Gogh Foundation in perpetuity. The largest collections of van Gogh art now exist in the Rijksmuseum (Statemuseum) Vincent van Gogh in Amsterdam, and the Rijksmuseum Kröller-Müller in Otterlo, also in Holland. The former contains the works donated by the van Gogh family, the latter has as its nucleus the collection of Mrs. Helene Kröller.

In more recent times three large retrospectives were organized according to major periods: *Van Gogh in Arles* (New York, 1984), *Van Gogh in Saint-Rémy and Auvers* (New York, 1986–87), and *Van Gogh à Paris* (Paris, 1988). The centenary of Vincent's death was celebrated in Holland in 1990 with massive representation from all periods of Vincent van Gogh paintings (shown in Amsterdam) and drawings (Otterlo). Given the rising and seemingly prohibitive insurance costs attending the mounting of such blockbuster shows, serious doubts have been raised as to whether their like will ever be seen again.

Thus the success of van Gogh was late in coming but remarkable in magnitude. There are indications from his letters of confidence in eventual recognition, but these are offset by as many or more doubts as he weighed the odds, struggled with illness, wrestled with financial circumstances, and wondered out loud if extant society would ever allow it. In the light of his

---

was then donated to Frankfurt City Gallery in 1911. At the end of 1937 it was confiscated by government authorities as "degenerate art" and sold for foreign capital to a private collector in New York. The painting sold by auction at Christie's in 1990 for a world record of $82.5 million.

286

posthumous success it is indeed both sobering and sustaining to recall the artist's modest comment from The Hague in 1882: "Smulders' workmen at the other store on the Laan saw the [lithographic] stone of the old man from the almshouse, and asked the printer if they could have a copy to hang on the wall. *No result of my work could please me better than that ordinary working people would hang such prints in their room or workshop*" (letter 245, my italics). Thus spoke the artistic evangelist and erstwhile religious pastoralist who generated, within a decade, a body of work which is remarkable for consistently high quality and because it can match quantitatively that of most artists with a normal, life-time career. The bulk of van Gogh paintings and drawings are now revered as masterpieces. What were the ingredients that permitted and nurtured this creativity?

Creativity

While the sciences may emphasize discovery of new facts of nature, arts the novel view, and performance the exquisite presentation, yet the most lauded participants within all three domains, the creative individuals, seem to use similar underlying processes. Creativity connotes inventiveness with value, and creative individuals are identified by their ability to formulate new relationships, no matter if the products be as superficially disparate as propositions, paintings, or poems. The products of creative endeavor may derive from newly discovered facts, observations, techniques or, equally important, they may spring from existing items. Work that we call creative is the bringing to fruition of something new, the casting of items into new relationships, the juxtaposing of words, pigments, or atoms into novel and useful arrangements.

Who are the authors of these creative works? Some were appreciated in their lifetimes. Others were recognized much later, sometimes with reluctant and belated praise. As Jonathan Swift remarked, "When a true genius appears in the world, you may know him by this sign, that the dunces are all in a confederacy against him." This dichotomy is an interesting subject, but we are less concerned with generating two lists than in identifying characteristics which distinguish outstanding from less creative people. And then we ask, how do they do it? Both questions are complex, and published analyses are sometimes enigmatic. Our focus remains Vincent van Gogh, so we will start with the artist's own thoughts on these subjects.

The philosophy of Vincent

There are many references to study and style, success and disappointment, improvement and frustration, within the correspondence of Vincent van Gogh. His propensity to engage in, "telling you the thoughts that come into my mind ... saying exactly what I mean" (letter 169), made many of his analyses both charming and revealing. The following quotations were extracted from letters; the operational headings are mine. They constitute a small selection of items which declare the artist's own philosophy.

1. Approach:
   "I do not know myself how I paint it. I sit down with a white board before the spot [scene] that strikes me; I look at what is before my eyes; I say to myself, 'that white board must become something' [yet] I come back dissatisfied. I put it away and when I have rested a little I go and look at it with a kind of fear. Then I am still dissatisfied because I still have that splendid scene too clearly in my mind to be satisfied with what I made of it. But I find in my work an echo of what struck me, after all. I see that nature has told me something, has spoken to me, and that I have put it down in shorthand. In my shorthand there may be words that cannot be deciphered, there may be mistakes or gaps; but there is [within it] something of what the glade or beach or figure has told me, and it is not the tame or conventional language derived from a studied manner or a system, but [rather a language derived] *from nature itself*" (my italics, letter 228, August 27, 1882).

2. Application:
   "Talent is long patience, and originality [is] an effort of will and of intense observation" (letter 470). This is a quote by Vincent from the novelist Gustave Flaubert (1821–1880).

3. Practice:
   "I work regularly on the *Cours de Dessin Bargue* [course in drawing by Bargue] and intend to finish it before I undertake anything else, for each day it makes my hand as well as my mind more supple and strong" (letter 136, 1880).

4. Feel for the subject:
   "I am getting an eye for this kind of country. ... If coming home with my canvas I say to myself, 'Look! I've got the very tones of old Cézanne!' I only mean that Cézanne like Zola is so absolutely part of the countryside, and knows it so intimately, that you must make the same calculations in

your head to arrive at the same tones. Of course if you saw them side by side, mine would hold their own, but there would be no resemblance" (letter 497).

5. Facility:
   "During your hasty visit did you see the portrait of Mme. Ginoux in black and yellow? That portrait was painted in three-quarters of an hour" (letter 573, January 23, 1889).

6. Color instinct:
   "I know for sure that I have an instinct for color ... [I know] that painting is in the very marrow of my bones" (letter 228, August 27, 1882).
   "I am absorbed in color – until now I have restrained myself, and I am not sorry for it"
   (letter 228, August 27, 1882).

7. Investment:
   "I consider making studies like sowing, and making pictures like reaping" (letter 233).

8. Signature and confidence:
   "the whole world can pronounce the name Vincent correctly ... they will surely recognize my work later on, and write about me when I'm dead and gone. I shall take care of that, if I can keep alive for some little time" (letter 435c) (quoted by Anton Kerssemakers).

These statements contain several important features: the concept of communing with nature (items 1, 4); evidence of intense concentration, hard work, and investment (items 2, 3, 7); embrace of repetitive exercises for learned skills (item 3, and see chapter 1); and some inherent and acquired skills (items 4, 5, 6). With regard to confidence (items 4, 8), the overall view must still be regarded as ambivalent; we need not list all the previously quoted letters expressing self-doubts and frustrations. However, the point that is worth making here is that all creative persons have at least moments of self-recognition. The remaining parameter in the equation for creativity is productivity (item 8). There are many talented individuals who produce virtually nothing; they are not called creative.

The term genius implies an exceptional and natural *capacity* for creative and original concepts, but not all geniuses are also productive. Galton[6] equated genius with accomplishment (quality, quantity, or even physical strength), which is certainly a minority position and is not taken here. The obverse of the talented dilettante is the high achiever who, in many cases at least, fails to break new ground and thus also fails to win the appellation of creative. It is of interest to ask about the predisposing factors for all of these categories, but perhaps a more important question relates to possible differences among creative individuals from different fields. We will start with some similarities between artists and scientists, but return later to possible differences.

The quintessential scientist, Michael Faraday

The contributions of Michael Faraday (1791–1867) to the theory of magnetism, electrochemistry, and electromagnetic induction, are well known scientific landmarks. His rise from the slums of London to the Royal Institution, and from assistant to Sir Humphry Davy to grand old man of British science, reads like a Cinderella story. Faraday preferred to be called a natural philosopher.[7] The following quotation reveals an underlying reverence of nature which has a strong analogy to Vincent's approach:

> "The philosopher should be a man willing to listen to every suggestion, but determined to judge for himself. He should not be biased by appearances; have no favourite hypothesis, be of no school, and in doctrine have no master. He should not be a respector of persons but of things. Truth should be his primary object. *If to these qualities be added industry, he may indeed go and hope to walk within the veil of the temple of nature*" (my italics).[8]

A personal point of view, and the courage to turn away from accepted views, are central to the approach of Faraday the scientist and of Vincent the artist. Interpreting and describing nature, and the necessity of hard work, are also common to both philosophies. This type of correspondence is remarkable among several of the most creative practitioners of art and science. Perhaps the major difference, if any, is that the artist may elect to go beyond nature, but nonetheless with reverence. This recalls Herbert Read's dictum for the artist, "his simple duty is, as Cézanne said, to be humble in the presence of nature or, as Vincent said, to be fearful in the presence of nature. *Then nature will speak through what is most natural in the artist, his sensations and feelings, his personal idea*" (my italics).[9]

The mechanisms of creativity

According to Kubie,[10] "the creative person is one who in some manner, which today is still accidental, has retained his capacity to use his preconscious functions more freely than is true of others who may potentially be equally gifted." William James (1842–1910), in *The Principles of Psychology,* called the preconscious the "fringes of conscious." Others use slightly different terminology, but they all place much emphasis on the involvement of the preconscious mind in the genesis of creative work.

Intellection and cogitation have become almost synonymous elsewhere, but there is a distinction that needs to be emphasized in the present context. The former refers to reasoning about things and testing items for reality and validity; this activity is associated with conscious pursuits; it is a large part of daily mental activity while awake. Cogitation, which has a Latin derivation connoting a shaking up and a falling out, refers more to ideas, memories, feelings and symbols. The result of cogitation is new combinations of these units; this is the realm of the preconscious. Facility in promoting the interaction between the preconscious and the conscious, i.e. cogitation plus intellection, is of vital import to creative work. Eiseley[11] agrees that "the freedom to create is somehow linked with the facility of access to those obscure regions below the conscious mind." He nonetheless reminds us that it is not the whole story, and he does not believe "that the time will come when each man can release his own Shakespeare."

Development and exploitation of the preconscious depend upon practice and experience. Take an acquisitional example: the conscious imitations of movements, facial twists, and sounds by infants become the acquired expressions, postures, and meaningful gestures which reside in the preconscious of the growing child. Or consider a more sophisticated process involving mental manipulations of data: different computational and transformational operations are given symbols and the mathematician calls upon these, both individually and in combinations, plays with them in the preconscious, and exhibits so-called intuitive thinking in solving problems. The activities of the preconscious are thus characteristically transacted in shorthand, which affords a much accelerated process. It has been established that the speed of preconscious processing may be drastically curtailed, and inventiveness may even be lost, if the otherwise creative individual is forced to verbalize the steps and justify every connection.

The unconscious is the home of infantile misconceptions and memories of traumatic experiences that interfere with "normal" activities. By definition these phobias, habits, and inhibitions are not readily perceived by the subject. It is the goal of the psychoanalyst to assist in their revelation. This may free the individual for more creative work, although, as Freud,[12] Storr,[13] and others have written, psychoanalysis per se will never explain creativity. Notwithstanding the promise of psychotherapy, many creative people have resisted the offer in fear that their talents are inextricably connected with their neuroses.[*] All parts of the neurological continuum, from the unconscious through the preconscious to the conscious, probably operate concurrently and the orchestration is supposedly vital to health. To be mired in the unconscious means unreality; to be anchored somewhere in the conscious means rigidity. The new combinations we associate with invention are rarely found by straining for them consciously, but rather by a process of free association involving the preconscious. The creative individual engages in a lively interaction between the preconscious and the conscious domains. Most importantly, productive creative thinking avoids interference from the unconscious.

The vulnerability of the preconscious to upset by the neurotogenic forces of the unconscious depends upon the extent to which the preconscious is utilized. Artists and scientists are just as susceptible to all the common ills of mankind, but because creative individuals employ the preconscious more than others, they are likely to be more susceptible to insults from the unconscious.[10] In neuroses, conflicts involving opposing wishes or ideas with an infantile or traumatic origin give rise to an unconscious anticipation of danger, which is experienced as anxiety. Notwithstanding the rather obvious consequence that a neurosis thus defined will inhibit creative thinking, a different and illogical connection between the two has often been stated, most notably starting with Lombroso,[14] whose book *Genius and Madness* was first published in 1864.

Creativity and mental illness

Cesare Lombroso (1835–1909), an Italian criminologist-anthropologist-psychiatrist who held professorial appointments at the Universities of Pavia and Turin, was convinced that genius was inherited and in some way connected with abnormality. Moreover, he was so impressed by the incidence of insanity in the families of his chosen subjects that he jumped to the conclusion that genius and insanity were co-inherited. Galton[6] is supposed to have rejected this hypothesis on statistical grounds, but his ponderous study took on a quaint aspect by including all manner of "geniuses" from "men of science" and "painters" to "oarsmen" and "wrestlers of the north country."

A study by Juda,[15] begun in 1927 and summarized in 1943, addressed the hereditary background and the physical and mental health of 294 geniuses (113 artists and 181 scientists) and their families, all from German-speaking

---

[*]   By neuroses we imply functional mental disorders in which reality testing is intact, in contradistinction to a psychosis in which reality testing is impaired.

regions. The average family size was six children and a remarkable number of the geniuses were first-born (38% of artists, 35% of scientists).[*] Only 10% of the primary subjects had no previously recognizable talent in their families, and there was a high number of intellectually prominent individuals among their children and grandchildren. That two-thirds of the artists and three-quarters of the scientists were mentally-healthy was taken as repudiation of the slogan "genius and insanity." However, the subjects and their families exhibited more psychoses and neuroses than the general population. The subclassifications indicated higher incidences of schizophrenia and manic depressive psychosis among the artists and scientists, respectively, but these are difficult to evaluate now because of changes in criteria.

Andreasen[16] evaluated 30 faculty members, over a 15-year period, at an American university workshop for creative writing. She claimed that the writers had a substantially higher rate of mental illness compared with 30 control subjects matched on sociodemographic grounds. A higher rate of affective disorders, especially manic depressive psychosis, was reported for the so-called creative group as well as their first-degree relatives. Jamison[17] reported that 38% of a British group consisting of 39 writers and 8 artists, which she deemed outstanding, had sought treatment for some form of affective disorder, especially manic depressive psychosis, compared with lifetime prevalence rates in that nation of about 6%. Her attempts to link hypomanic episodes and seasonal mood swings with productivity were unconvincing. Rothenberg[18] severely criticized both reports[16,17] on the grounds that little consideration was given to the subjects' reasons for participating in the studies, and the criteria for judging them creative was left unexplained. Furthermore, Andreason's[16] self-reliance on evaluation of relative mental health was potentially biased because the subjects and controls were already known to her; Jamison[17] rested her case on the subjects' own reports of seeking medical treatment.

Antonin Artaud was hardly in the same league as the creative people we are discussing, but his case is worth mentioning because he insinuated himself into the van Gogh story by his attack on Dr. Gachet. Artaud was critical of his own treatment at Rodez, the final institution to which he was committed, and in much of his writing he chastises and rebukes doctors generally, and psychiatrists in particular, for what he calls their atavistic reflex to become the innate enemy of all geniuses.[19] Paradoxically, it was in

---

[*]   The observation of a relatively high rate of achievement in first-borns has been repeated many times since.

Dr. Ferdière's clinic that Artaud, who had been silent for years, began to write again and also to paint and to draw.

The consensus of serious studies suggests that there is no cause and effect relationship between insanity and creativity, but that some forms of mental illness are not incompatible with creativity in the long run. Moreover, there is no indication for sickness being a prerequisite. On the contrary, frank or masked neurotic tendencies are the bane of creativity. The productivity of scientists and artists with episodic illness is usually associated with their periods of wellness. A prime example is Vincent van Gogh who was so productive in the lucid intervals between attacks but was completely incapacitated, unable to paint or to write, during most of his medical crises. It was his commitment to making a contribution in art that brought Vincent back to his work, after every insult to his health.

Commitment and aggrandizement

Scientists and artists who have made a contribution emphasize the importance of making a commitment to the task. They *want* to create something. Vincent talked about working hard at the basics of drawing, perspective, and color-theory as well as the day-to-day production of canvases. Faraday called it industry. Thomas Edison said that genius was 1% inspiration and 99% perspiration (newspaper interview, 1931). Basic to the concept of industry is meaningful application with focus and organization; Vincent may have led one of the most chaotic lives in nineteenth century Europe, but he certainly was organized and focused in his art work during those last ten years, and he was motivated to paint. Rothenberg[18] feels that the most persistent characteristic of creative people is motivation. Kubie[10] observed that "among those who strive to be creative, the sense of dedication is always present, no matter how it is masked by matter-of-fact and workaday attitudes." Along the same lines, Sternberg and Lubart[20] included the willingness to surmount obstacles, to persevere, and to avoid derailment as crucial to success. The depth and intensity of the committed scientist or artist can sometimes be startling to the uninitiated and some of Vincent's comments, for example, especially if taken out of context, look like self-aggrandizement.

Kubie[10] is convinced that "all activities which men think of as *vocations* or *callings* have a subtle flavor of discreet megalomania, a megalomania masked by outward humility, poverty, chastity, religious devotion, or dedication to a life of scientific research." Van Gogh certainly felt called to be an artist: "it is my belief that it is actually one's *duty* to paint the rich and magnificent aspects of nature" (letter W7). In chapters 2 and 9, I have discussed the temporal and philosophical aspects of his turn from an embrace of religion to a reverence of nature,[*] and this is the crux of Kodera's book.[21] There are repeated references in Vincent's correspondence to indicate that the new direction was nonetheless accommodated by his existing intellectual tools. The approaches were similar, and it is not without substance that D'Evelyn[22] remarked upon van Gogh's intuitively theological approach.

The artist and society

To be creative in any field takes enormous amounts of energy and time devoted specifically to that field. The individual who wishes to make a contribution may find numerous social pressures acting as roadblocks. Hayes[23] observed that if your society values intellectual activities, encourages you to be interested in intellectual activities, encourages you to believe that you can succeed in them, helps you to obtain the necessary education, and does not impose other occupations on you which preempt your time, then your chances of becoming a creative person will be better than average. Most people can identify social pressures which have helped or hindered their pursuits toward creative work. Also, if successful artists or scientists do carve out sufficient time for the creative task they are often judged by contemporary society to be withdrawn, self-centered, and preoccupied with only a single mission. Subsequent generations may acknowledge this same focus upon the task, to the exclusion of everything else, as wholesome commitment.

One episode among many for Vincent van Gogh concerned his reluctance to take the time to help his mother and sister move from Nuenen to Breda, after his father's death and when the grace period on the vicarage had expired. From Theo's perspective Vincent's refusal appeared oafish, and coupled with Vincent's arrival on his doorstep in Paris shortly thereafter (and two months prematurely), was doubly difficult to excuse. From Vincent's perspective both responses were simply extensions of his commitment to art and a vigorous rejection of anything that interfered. Several past and contemporary Nobel Prize – winning scientists have remarked on their ability to concentrate on a single problem for an extended period of time (hours, days or even weeks compared with the seconds or minutes of the average person). This is the origin of the misnomer "absent-minded profes-

---

* In the 1880's, van Gogh, Monet, Gauguin, and Cézannne all developed a skepticism about industrial civilization. Their faith in nature, places that were not urban, and people who were not urbane, largely replaced their religion.

sor" for an individual whose only "absence" is from outside distraction. If any other unusual activity follows, then the accusation of abnormality is not far behind, because as Ober[24] puts it, "the plural of anecdote is data." Accordingly, creative people often find themselves burdened with exotic attributes which they either live with, or down, according to their personality. In the extreme, they feel so out of the mainstream that they have doubts about their own sanity.

The artist finds himself in the paradoxical situation of being encouraged by peers and critics to go to the limits of nature, and even beyond. Julien Levy[25] said, "the artist alone among all the world has the duty to pursue a special point of view to its farthest reach, to exaggerate and embellish just the things which others prudently modify, diminish, or retrace toward the common, less lonely, comprehensible, and useful center." By embracing this approach the creative artist puts himself more at odds with society than does the creative scientist.

Medawar[26] made a comparison: "if a scientist were to cut off his ear, no one would interpret such an action as evidence of an unhappy torment of creativity." Notwithstanding the poignancy (as well as the humor) of this statement, none of the medical crises of Vincent, including the bizarre ear-cutting affair, can be construed as having a positive effect on that artist's creative potential. However, Medawar's quip correctly implies that a significant fraction of the population at large does associate "abnormal" behavior with artists. Moreover, certain exotic behaviors are tolerated and anticipated in artists, and may even encourage some of them to adopt such a stereotypic lifestyle while awaiting recognition. It sometimes becomes a habit.

Creative persons are not alone in defending themselves against change, but their reluctance receives special support from culturally accepted rationalizations. We have already attempted to dispel the myth that neuroses or even psychoses are inextricably intertwined with the capacity for creativity. The reluctance of some creative artists to undergo therapy for mental illness, alcoholism, or other substance-abuse out of fear of losing their creativity is unfounded, but is in no small part sustained by public image. Reports of a causal relationship between affective disorder and creativity, for instance, or signs of insanity in paintings, are often patently misquoted or seized upon with an unfortunate lack of organized skepticism. These irresponsible commentators confound societal perception, and might better remark on the great number of creative individuals who have no apparent disease.

Freud proposed that artistic creation is a sublimation of unconscious processes, and his disciples have sustained the notion. Unfortunately, "the assumption that all infantile conflicts can be solved, or ought to be solved, and that all man's emotional problems are dissipated by the whirlwind release of repeated, regular orgasm, dies hard."[27] Nagera[28] felt that because van Gogh had abandoned all thoughts of women, marriage, and a family he was able to produce instead a very large number of paintings during the last two and a half years of his life. This florid hypothesis pales before the simple declaration by Vincent that he wanted to paint, and his demonstrable commitment to being an artist.

Art therapy and work therapy; Vincent's case

A corollary to sublimation, namely that artistic activity can be therapeutic, arose from the Freudian background. Art therapy may be one useful ingredient in the treatment of manic depressive psychosis or schizophrenia, but there is no evidence that this alone will ever cure mental illness.[10,18] In any event I have argued against these disorders in Vincent's case, in favor of a toxic psychosis, which is not amenable to art therapy. Vincent van Gogh neither painted himself into the asylum nor worked his way out of it.

Van Gogh voluntarily entered St. Rémy asylum because he had already suffered two terrifying episodes of psychotic illness and, given one experience of complaint by the townspeople of Arles, he feared compulsory incarceration. The intermittent nature of his illness allowed him to be highly creative during the periods between crises. The superintendent, Dr. Peyron, permitted him to paint, gave extra space within the institution, and provided supervision in the field. At the time of Vincent's exit from St. Rémy asylum, Dr. Peyron wrote that Vincent was "cured." Subsequent events proved otherwise. The director's reluctance to fully endorse the move or to have Vincent travel unattended were noted in chapter 6, and may be compared with the inanity of the final evaluation.

Dr. Rey in Arles and Dr. Gachet in Auvers were even more enthusiastic than Peyron about work therapy for Vincent. They have been commended for this, and rightly so. My point is not that this type of activity is unreasonable but that it will rarely be enough. In the case of acute intermittent porphyria the current therapy involves symptomatic treatment during crises, recommending a high carbohydrate and adequate caloric diet, and counseling against the use of certain drugs and alcohol. This was unknown in the nineteenth century.

It has been noted that Dr. Rey gave Vincent good advice about eating properly, and admonished him for drinking and smoking too much. Less is known about the recommendations of Drs. Rivet and Gruby in Paris, although they both seem to have been against excess of alcoholic beverages. Dr. Gachet made some effort to introduce Vincent to three-course meals in his home, although the patient resisted. Anecdotal evidence suggests that

Dr. Gachet also gave some thought about the artist's overexposure to turpentine vapor. Whether these gentlemen were insightful, intuitive, or simply commonsensical about Vincent's illness is beside the point. Unfortunately, van Gogh rarely followed their advice and some of his crises were undoubtedly influenced, if not directly provoked, by periods of fasting and malnutrition and drinking absinthe or other alcoholic beverages.

Alcohol and other substance abuse

Many creative people have been alcoholics. All fields are represented and Rothenberg[18] provided lists. A significant number of prize-winning writers had their careers curtailed and their lives shortened by alcohol, although they did their best work while sober.[29] This is in keeping with our present understanding, namely that creative thinking and creative work occur when the creator is lucid, not intoxicated, and not psychotic. This is a consistent finding regardless of individual susceptibilities to mental illness or proclivities for alcohol and other drugs.

Quite often there are secondary effects of alcoholism involving some relationship with psychiatric disorders. While the toxic effects of excessive alcohol strike primarily the major organs, especially the liver and kidney, they are also manifested as changes in psychological and social functioning. Alcoholism can mask, mimic, precipitate, or independently coexist with a gamut of psychiatric disorders.[30] Some patients use alcohol to medicate painful affects and disturbing psychiatric disorders to the extent that alcohol dependence becomes the new primary problem. Suicide occurs at a higher rate among alcoholics than in the general population. All of these types of interaction have been encountered among creative people, but again we must add that there are many who have been equally successful and showed no signs of addiction to alcohol, thus negating the concept of a concomitant susceptibility to the charms of ethanol. Also, there is no indication that alcohol or other drugs will nurture creativity in the unprepared mind.

Alcohol and other drugs such as marijuana, mescalin, the ergot alkaloids, opiates, and barbiturates enhance, distort, or even eliminate normal information received by cells that control the senses. These drugs also act at centers of the brain where moods and emotions are regulated. They can produce feelings of well-being and euphoria as well as paranoia, fear, and depression. A high dose of any of these drugs on the nervous system may cause a stimulant psychosis, a form of toxic psychosis which is characterized by extreme paranoia and is often accompanied by violence and injury.[31] Repeated drug use leads to a compulsive urge to continue using the drug, a tendency to increase the dosage, and physical and/or psychological dependence. By dependence we imply an adaptation of the body to the presence of a drug such that its absence produces withdrawal symptoms.

For most of this century, the term "addiction" was used to convey the idea of a drug user's physiological adaptation to psychoactive substances. The withdrawal symptom was thought to be a key element in maintaining addictive behavior, in other words the addict went back to the drug to relieve the pangs of withdrawal. Today the term is used in a much broader sense. The present key concept in addiction, at least colloquially speaking, is compulsive use; the addicted person continues to use the drug in spite of knowing that the drug is causing problems. The 1987 revision of the *Diagnostic and Statistical Manual of Mental Disorders* [DSM – IIIR], avoids the use of the term addiction, relies on a Chinese menu approach by requiring three out of nine criteria to deem a patient to be drug-dependent. Manifestation of tolerance and exhibition of withdrawal symptoms are two criteria: the rest of the criteria such as spending a great deal of time acquiring the drug, and interference with work or home activities, are behavioral.

Vincent and alcohol

In chapter 3, I collected all of the comments by Vincent, his doctors, and friends on his drinking problems. Perhaps the most appropriate of these in the present context is Vincent's statement to Theo, just two months after arrival in the asylum at St. Rémy:

"I [now] live soberly [here in St. Rémy] because I have a chance to, I drank in the past because I did not quite know how to do otherwise. Anyway, I don't care in the least!!! Very deliberate sobriety – it's true – leads nevertheless to a condition in which thoughts, if you have any, move more readily. In short, it is [the] difference [between] painting in gray or in colors. I am [now] going to paint more in gray, in fact. ... I am well, however, and I have a feeling rather like I had when I was younger, when I was very sober, too sober they used to say then, I think" (letter 599, July 5, 1889).

My interpretation is that van Gogh was to some extent blaming society both in the opening remarks, i.e. 'to drink was the thing to do' and in the close 'things were not so great even before the drinking problem.' His protest about not caring is all too typical of someone with a problem. The remark about sobriety being less 'colorful' brings us to our next, and possibly the most intriguing consideration, namely that many creative artists suppose that they need the stimulus of alcohol or other drugs to open their brains to a new view of the world. The example of mescaline is well documented.

"The Doors of Perception"

In 1886, the German pharmacologist Louis Lewin initiated a series of studies on a species of cactus which was subsequently named for him, *Anhalonium lewinii*. It was new to Western science but not to the natives of old Mexico. They were well aware of the psychological effects resulting from eating the root which they called peyote. The active principle was soon identified as mescaline (3,4,5-trimethoxybenzeneethanamine), a relatively simple alkaloid. According to the standards of an earlier day, experimentalists gave themselves at first small and then increasing doses of mescaline, and confirmed that profound changes in consciousness were induced.

Aldous Huxley[32] was not the first to report, but his experiences have been the most widely quoted. He likened his own trial with mescaline to opening the "doors of perception," and went on to recommend its wider usage for "an occasional trip through some *chemical door in the wall* into the world of transcendental experience." In what now appears as a warning, but was then more of a promise, Huxley notes that "the man who comes back from the door in the wall will never be quite the same man who went out." Elsewhere,[33] Huxley enjoined aspiring artists, visionaries, and mystics to seek technical help from biochemistry, pharmacology, physiology, psychiatry, and psychology. The scientific specialists were also asked to turn out from their respective pigeonholes and assist in opening the *chemical door*. This is a position that has largely fallen out of favor. However, before we cast too many stones at Huxley, recall that Thomas Edison and Jules Verne endorsed cocaine-containing *Vin Mariani* at the turn of the century. Notwithstanding all of the potentially deleterious consequences of abuse of alcohol and other drugs, their connection with the creative process remains stubbornly and titillatingly intact.

Rothenberg[18] suggested that the use of alcohol by creative writers derives from a necessity to cope with anxiety. The revelation of unconscious, stressful data is considered a rather inevitable spin-off to the creative process[10,18] and the writer may resort to the bottle as a sedative. Kubie[34] mentions alcohol, drugs that affect emotional processes, and toxic-delirious insults among the many relevant influences on the to-and-fro between immature and mature levels of symbolic function, which is important to the creative process. All of these authors stress that the effects will depend on the pre-existing mental and emotional background of the subject.

Schneider[35] deduced that Coleridge's opium habit would not have been sufficient alone to induce the fantastic dreams and confer the imaginative stimulus to compose "Kubla Khan" and other masterpieces. Hayter[36] surveyed eighteenth and nineteenth century writers and was unable to discern a pattern of influence for opium: "opium works on what is already in a man's mind and memory, and what was already there in the eight writers [De Quincey, Coleridge, Poe, Baudelaire, Crabbe, Collins, Thompson, and Keats] … was extremely varied." In a new approach, Monroe[37] suggests that bursts of electrical activity deep within the limbic system, "brainstorms" as he calls them, are closely associated with inspirational thought, and that this region of the brain is exquisitely sensitive to many drugs. Monroe has interesting chapters on Vincent van Gogh and August Strindberg and their proclivity for absinthe. Other commentators seem to have a Victorian reluctance to confront such matters.

It has been my experience that certain topics such as alcohol, drinking habits, and absinthe are often excluded from indices, even from those which serve biographies giving reasonable attention to these subjects in the text. My first thought, that this was simply an oversight which happened to be inconvenient to my research, is now replaced by the suggestion that these subjects are not only under-reported but often inaccessible. A case in point is the very readable and otherwise informative collection of letters and comments on the poet Ernest Dowson by Fowler and Maas,[38] which of course includes the famous pun, "I understand that absinthe makes the tart grow fonder." They also spend several paragraphs in the narrative on the poet's drinking habit, effects on his health, and early demise, but list nothing in the index to guide the reader to the absinthe pun or related subjects.

Vincent van Gogh and absinthe

I have purposefully avoided saying that Vincent van Gogh was addicted to absinthe. The current psychiatric definition makes it even more inappropriate. His condition is best described as having a fondness for the liqueur, exhibiting a proclivity for absinthe drinking, and suffering from malnutrition because he took too many of his calories that way. The deleterious effects of excessive consumption of absinthe were attested to in chapter 3, and the mood swings and hallucinations associated with absinthism have been documented. Absinthe drinking was a popular activity for many in nineteenth century France. However, it is very difficult to judge the relative exposure of Vincent compared with a "typical" citizen such as the postman Roulin (Figure 11.2).

I am impressed by the assessment of Dr. Rey that Vincent ate poorly during days in the field and then often drank alcoholic beverages when he

Figure 11.2. Vincent van Gogh: *Portrait of Joseph Roulin*, 1888, Arles, pencil, quill and reed pen and reed pen and brown ink, on wove paper, 31.5 × 24.0 cm (12.4 × 9.4″), J. Paul Getty Museum, Malibu, California.

was tired at night, rather than taking the time or energy to cook. At other times, in the company of Toulouse-Lautrec in Paris or Paul Gauguin in Arles, for example, anecdotes suggest that he may have engaged in some bouts of drinking. There are only innuendoes about how much absinthe he drank during the evenings that he painted *The Night Café* on location, but it would be more than surprising if he did not have at least a few glasses. During the week-ends when he left the asylum at St. Rémy and visited Arles he always sought out old friends and most likely drank. More importantly, there is a plethora of comments from his doctors, colleagues, family, and the artist himself, to indicate that at times he drank too much.

How much Vincent's drinking affected his view of the world at that time, and how much those visions influenced his later renditions on canvas, are difficult to ascertain, but no serious scholar would exclude the possibility. On the other hand, the effects on his underlying illness are more obvious. Alcohol can provoke medical crises in patients with acute intermittent porphyria, and we have shown that at least one terpene in absinthe, thujone, will upset the heme pathway, which is the locus of vulnerability in this disease. The mechanism exists for a greatly increased sensitivity to absinthe in individuals with this rare disease.

The van Gogh legacy

"It is possible that these great geniuses [Rembrandt, Delacroix, Zola, Balzac, Millet] are only madmen, and that one must be mad oneself to have boundless faith in them and a boundless admiration for them. If this is true I should prefer my insanity to the sanity of others" (letter B13). Thus wrote Vincent to Émile Bernard, toward the end of July 1888. It was five months before his first major medical crisis. The language was exaggerated and so was his art. For comments such as this, bizarre acts, exotic behaviors, and a seemingly wretched life ending in suicide, he captured the attention of even the blasé majority of succeeding generations.

Vincent was not mad. He achieved much in the face of adversity, and his artistic masterpieces exemplify the uplifting conclusion that creative work can be compatible with underlying debilitating illness. His life style exacerbated his condition and compounded his problems.

Vincent van Gogh's legacy for future artists, and all who would be creative, springs from his ample demonstration of hard work in the basics of technique and theory; a receptivity toward what others have thought and achieved; an organized skepticism; a willingness to take chances; unwavering commitment to his vocation; a reverence of nature; and a decision to undertake a valuable task and to work toward a meaningful goal. To all this were added intelligence and ability, which were inherited, as was his illness.

References and notes

1. Johnson HA. 1934. No Madman. *The Art Digest* **8**: 11.
2. van Gogh VW. 1978. Some additional notes to the memoir of Vincent van Gogh. pp LIV–LXVII. in: *The Complete Letters of Vincent van Gogh.* 2nd ed. Boston: New York Graphic Society.
3. Rewald J. 1986. The posthumous fate of Vincent van Gogh 1890–1970. pp 244–254, in: *Studies in Post-Impressionism.* I Gordon & F Weitzenhoffer (eds.). New York: Harry N. Abrams Inc.
4. Rewald J. 1978. *Post-Impressionism.* 3rd ed. New York: The Museum of Modern Art.

5. Reese KM. 1990. Brief history of a painting by van Gogh. *Chemical and Engineering News* **68**: 56.

6. Galton F. 1952. *Hereditary Genius: an Inquiry into its Laws and Consequences*. New York: Horizon Press. First published in 1869, London: MacMillan.

7. Agassi J. 1971. *Faraday as a Natural Philosopher*. Chicago and London: University of Chicago Press.

8. Faraday M. 1816. Lectures to the City Philosophical Society on Physics and Chemistry. (see: Bence-Jones H. 1870. *The Life and Letters of Faraday*. vol 1, p 220, quoted in reference 7).

9. Read H. 1967. *Art and Alienation: The Role of the Artist in Society*. New York: Horizon Press.

10. Kubie LS. 1958. *Neurotic Distortion of the Creative Process*. Lawrence: University of Kansas Press.

11. Eiseley L. 1962. *The Mind as Nature*. New York: Harper and Row.

12. Freud S. 1959. An Autobiographical Study. vol 22. in: the standard edition of the *Complete Psychological Works of Sigmund Freud*. J Strachey et al. (eds.). London: The Hogarth Press and the Institute of Psycho-Analysis.

13. Storr A. 1988. *Churchill's Black Dog, Kafka's Mice, and other Phenomena of the Human Mind*. New York: Grove Press.

14. Lombroso C. 1887. *Genie und Irrsinn in ihren Beziehungen zum Gesetz, zur Kritik und zur Geschichte. Von C. Lombroso. Mit Bewilligung des Verfassers nach der 4. Aufl. des italienischen Originaltextes ubers. von A. Courth*. Leipzig: P. Reclam. (See also: Lombroso C. 1891. *The Man of Genius*. London: Walter Scott.)

15. Juda A. 1949. The relationship between the highest mental capacity and psychic abnormalities. *American Journal of Psychiatry* **106**: 296–307. (This is an abridged, English edition of a German text.)

16. Andreasen NC. 1987. Creativity and mental illness: prevalence rates in writers and their first-degree relatives. *American Journal of Psychiatry* **144**: 1288–1292.

17. Jamison KR. 1989. Mood disorders and patterns of creativity in British writers and artists. *Psychiatry* **52**: 125–134.

18. Rothenberg A. 1990. *Creativity and Madness, New Findings and old Stereotypes*. Baltimore and London: The Johns Hopkins University Press.

19. Greene N. 1970. *Antonin Artaud, Poet without Words*. New York: Simon & Schuster.

20. Sternberg RJ & Lubart TI. 1991. An investment theory of creativity and its development. *Human Development* **34**: 1–31.

21. Kôdera T. 1990. *Vincent van Gogh: Christianity versus Nature*. Amsterdam and Philadelphia: John Benjamins Publishing Company.

22. D'Evelyn T. 1990. Van Gogh's capacity to surprise. *The Christian Science Monitor* p 14, September 6, 1990.

23. Hayes JR. 1981. *The Complete Problem Solver*. Philadelphia: Franklin Institute Press.

24. Ober WB. 1991. The man in the scarlet cloak: the mysterious death of Peter Anthony Motteux. *The American Journal of Forensic Medicine and Pathology* **12**: 255–261.

25. Levy J. 1977. *Memoir of an Art Gallery*. New York: G.P. Putnam's Sons.

26. Medawar P. 1979. *Advice to a Young Scientist*. New York: Basic Books Inc.

27. Storr A. 1972. *The Dynamics of Creation*. New York: Atheneum.

28. Nagera H. 1967. *Vincent van Gogh*. London: Allen & Unwin Ltd.

29. Goodwin DW. 1988. *Alcohol and the Writer*. Kansas City & New York: Andrews & McMeel.

30. Solomon J. 1989. Alcoholism and psychiatric disorders. chapter 9, in: *Alcoholism: Biomedical and Genetic Aspects.* HW Goedde & DP Agarwal (eds.) New York: Pergamon Press.
31. McLellan T, Bragg A. & Cacciola J. 1986. *Escape from Anxiety and Stress.* New York: Chelsea House Publishers.
32. Huxley A. 1970. *The Doors of Perception.* New York: Harper & Row.
33. Huxley A. 1955. *Heaven and Hell.* New York: Harper & Brothers.
34. Kubie LS. 1945. The value of induced disassociated states in the therapeutic process. *Proceedings of the Royal Society of Medicine* **38**: 681–683.
35. Schneider EW. 1953. *Coleridge, Opium, and Kubla Khan.* Chicago: University of Chicago Press.
36. Hayter A. 1968. *Opium and the Romantic Imagination.* Berkeley & Los Angeles: University of California Press
37. Monroe RR. 1992. *Creative Brainstorms: The Relationship between Madness and Genius.* New York: Irvington Publishers, Inc.
38. Flower D. & Maas H. 1967. *The Letters of Ernest Dowson.* Cranbury, New Jersey: Associated University Presses Inc.

Chapter 12
Summary

> *Life is not long for anybody, and the problem is
> only to make something of it.*
> Vincent to Theo, letter 397, from
> Nuenen, April, 1885.

Vincent van Gogh was a wonderfully accomplished artist whose work is now widely appreciated. He created a great number of masterpiece paintings and drawings in just one decade devoted to art. His productivity is even more remarkable when considered in the context of his debilitating illness. During his last two years he suffered from medical crises that were devastating, but in the intervening periods he was both lucid and creative. He left a profound, soul-searching description of his jagged life in his correspondence, which provides the basis for the present analysis.

A careful review of data from the artist's letters and other contemporary sources indicates that Vincent suffered from an inherited disorder manifested by severe and manifold neurological problems, ranging from gastrointestinal pains to fits with hallucinations. His condition was exacerbated by his life style, which was marked by inadequate nutrition, abuse of alcoholic beverages, chronic smoking, environmental exposure, and the development of an abnormal affinity (pica) for terpenes. The intermittent nature of his illness, the sudden onset of crises, and the rapid return to normalcy after each episode, are all notable. The gamut of symptoms is best explained by a toxic psychosis. Within that category, the disease entity which most closely fits all of the data is acute intermittent porphyria [AIP], which was adopted as a working hypothesis for Vincent's underlying illness. This retrospective diagnosis has been compared and contrasted with other suggestions in the literature. AIP was not understood in Vincent's time; even today it tends to be under-diagnosed.

The case for AIP is strengthened in many ways. It is well documented that infections, fasting and malnutrition, alcohol, and several drugs can precipitate crises. On the other hand, some carriers for the defective gene in AIP avoid all medical problems by leading a careful and sober life-style. Vincent's unrelenting pursuit of his artistic goals, often at the expense of his "daily bread," provoked his malady. The van Gogh family history of mental illness provides peripheral support for the hypothesis of an inherited disorder.

Sensitivity to alcoholic beverages is known for patients with AIP, and an acute response to absinthe in particular would be reasonably anticipated because it was high in both alcohol and essential oils. Our recent data indicates a porphyrogenic potential for thujone (the toxic principle of absinthe), camphor (a nineteenth century medicament), and pinene (the major constituent of turpentine); these are the compounds associated with van Gogh's pica for terpenes.

For over a hundred years it has been known that absinthe abuse alone can induce fits with hallucinations, seizures, and gastrointestinal upset due primarily to terpenes in the liqueur, especially thujone. Vincent's proclivity for absinthe should be tempered by the context of its widespread popularity in France, especially in Paris and Arles, during his lifetime. The important point here is that a disease such as acute porphyria would place the individual at increased risk after ingestion of absinthe.

The chemistry of absinthe provides an interesting stage for viewing the great advances in organic chemistry that occurred during the latter decades of the nineteenth century. There were other positive ramifications such as the application of thujone and its chemical cousin camphor for experimentally induced seizures which mimic epilepsy. These were important scientific developments during the beginning of the twentieth century. My chapter devoted to Vincent's doctors looks back at a difficult and changing time, with the beginning to a chemical underpinning of medical understanding and the start of rational drug therapy. A proper realization of the potential effects of ingested chemicals on the brain as well as other major organs was finally breaking through.

A palette which favored the red end of the spectrum, and included much yellow and green, was employed by Vincent in his so-called "high yellow" paintings. He made the interesting observation that his illness had possibly helped him reach a "high yellow note." These paintings were significant in number; they were produced in multiple locations (from Paris through Arles and St. Rémy to Auvers-sur-Oise); and their execution was episodic rather than sustained. Chemical and physical insults were explored as possible causes of a temporary xanthopsia, an abnormal yellow vision that might inspire an unusual palette after return to normalcy. The most plausible chemical candidate is santonin, a sesquiterpene which was widely used as an anthelminthic in the nineteenth century, and overdosage is known to cause yellow vision. It was recommended in Vincent's sourcebook of home remedies, Raspail's *Manuel Annuaire de la Santé*, but in the absence of direct evidence that the artist overdosed on the drug, the hypothesis of modified perception was rejected in favor of artistic preference. Nonetheless, the concept of drugs, including alcohol, affecting an artist's "view of the world" is far from dismissed by this particular example.

Van Gogh's ear-cutting affair has been reassessed in view of the likelihood that he was undergoing a toxic psychosis. The possibility of accident is entertained here as well as in connection with his suicide. The interval between Vincent's recovery from his last medical crisis in St. Rémy and the self-inflicted gunshot wound in Auvers is certainly in accord with his anticipated, and declared, safety period of three months. My emphasis in both the ear-cutting episode and the eventual suicide is upon the importance of the underlying illness rather than psychoanalytic explanations.

In discussing the subject of creativity I have compared Vincent's philosophy with that of the physicist Michael Faraday, and noted extensive similarities. Both emphasized hard work and a reverence of nature. Artists and scientists are popularly perceived as having eccentric tendencies. However, society seems to tolerate and even encourage exotic behavior in artists more readily than in other creative individuals. The bizarre and unfortunate acts that are associated with Vincent van Gogh have undoubtedly contributed to his international and cross-cultural appeal by eliciting sympathy and provoking titillation. However, the appellation of mad does not apply. There is no evidence that mental illness is either a prerequisite to, or an inevitable consequence of, creative work.

Vincent suffered from a debilitating disease and his hardships were in major part due to his underlying illness. In spite of everything he suffered, he was creative and productive. His life was not long, but indeed he was able to "make something of it."

Primary publications

Some of the subjects of this book have been addressed in journal articles by the author and his colleagues. The background information and the chemical and biological justifications on these topics were generally expanded in the present format, but it was neither appropriate nor worthwhile to repeat some of the published data. Each source has been cited in the appropriate chapters. They are assembled here for the convenience of readers who may wish to consult the original papers.

Arnold WN. 1988. Vincent van Gogh and the thujone connection. *Journal of the American Medical Association* **260**: 3042–3044.

Arnold WN. 1989. Absinthe. *Scientific American* **260**: 112–117.

Arnold WN, Dalton TP, Loftus LS & Conan PA. 1991. A search for santonin in *Artemisia pontica,* the other wormwood of old absinthe. *Journal of Chemical Education.* **68**: 27–28.

Arnold WN & Loftus LS. 1991. Xanthopsia and van Gogh's yellow palette. *Eye* **5**: 503–510.

Loftus LS & Arnold WN. 1991. Vincent van Gogh's illness: acute intermittent porphyria? *British Medical Journal* **303**: 1589–1591.

Bonkovsky HL, Cable EE, Cable JW, Donohue SE, White EC, Greene YJ, Lamprecht RW, Srivastava KK, & Arnold WN. 1992. Porphyrogenic properties of the terpenes – camphor, pinene, and thujone; with a note on the illness of Vincent van Gogh. *Biochemical Pharmacology* **43**: 2359–2368.

The thuja tree that grew on van Gogh's grave for fifteen years. It is now growing in the garden of Dr. Gachet's home, owned by Colonel Gilbert Vandenbroucke (standing at the entrance to the three storied house) and Ursula Vandenbroucke. Photograph courtesy of Alain Mothe.

Afterword
Burial, Reburial, and the Thuja Tree

Dr. Gachet placed a small ornamental tree on the first grave of Vincent van Gogh. The initial burial arrangement was for a fifteen year, nonrenewable concession, and on June 9, 1905, a new and bigger site was arranged by Johanna van Gogh-Bonger, to be held in perpetuity. On June 15, 1905, Vincent's remains were duly transferred. Dr. Gachet's son recalled how the roots of that ornamental tree had completely entwined the casket, and it was indeed remarkable that the shrub survived extrication and transplantation to the Gachet garden.

When I first saw that ornamental tree it was almost one hundred years old and almost as tall as the three-storied house. It is a thuja tree, *Thuja occidentalis,* source of thuja oil and namesake of thujone. The most toxic constituent in absinthe was thujone.[*]

Vincent was enamored of cypress trees in the south of France and he incorporated symbolic, flame-like renditions of them in several canvases from St. Rémy. The cypress does not grow as far north as Auvers, but Vincent maintained the same style and symbolism when he painted the ornamental trees in Dr. Gachet's garden. This had undoubtedly inspired the choice of grave decoration. It was indeed an unwittingly pathetic choice to have a thujone-bearing plant on his mortal remains for fifteen years.

---

[*] Arnold WN. 1988. Vincent van Gogh and the thujone connection. *Journal of the American Medical Association* **260**: 3042–3044.

Acknowledgments

Special thanks are due to my colleague Loretta Loftus for sharing ideas about the illness of Vincent van Gogh, making her extensive library available to me, and checking quotations. I appreciate her enthusiasm for the subject and her help with this book.

Louise Arnold, Roy Baynes, Allan Cooke, Robert Garrison, Chi-Wan Lai, Brian Lawrence, Loretta Loftus, Stuart Munro, Manuel Pardo, Ken Schmitz, Barry Skikne, Jill Warnock, and Fred Whitehead offered useful suggestions on one or more drafts of various chapters. Several scholars gave expert technical advice, materials, or both; they include Conrad Arnold, Herbert Bonkovsky, Blair Bowers, Bruce Bradley, A.J. Collet, John Doull, Remo Fabbri Jr., R. Gelius, Jessie Gifford, Carla Green, Jan Hellings, Jessica Hellings, John Kepes, Jemshed Khan, James Kutney, Brian Lawrence, Russell Monroe, William Ober, Henry Peters, Mike Pronko, James Ravin, Andrew Salter, Fred Samson, Charles Sittler, Jack Valdovinos, Lutz Weber, Robert Wiles, and Burt Zerner.

My thanks are extended to Patrick Conan, Nadine Dormoy, Christian Ferri, Eveline Guého, Claude Millon, Alain Mothe, and Gilbert and Ursula Vandenbroucke who assisted in France; as did Jan Langeveld, Robert Naborn, Cindy Pronko, and Paul Wilson in Holland; Barry Artist and Hubert Britton in England; and Brita Velghe in Belgium. The Kansas City artists Jim Gubar, Lynn Huber, Hugh Merrill, Jane Pronko, Mike Stack, and David Strout participated in the studies with yellow filters.

The editorial acumen of George Adelman, as well as his encouragement and sustained interest in this project, are greatly appreciated. The expertise of Dorothée Engel in the production of both the English and the German editions of this book is gratefully acknowledged.

List of Figures and Tables

314

Name Index

The relatives of Vincent van Gogh are in **boldface**.
Abbreviations: b. = born, c. = circa (about), d.= died, fl. = flourished.

Anquetin, Louis [artist] (1861–1932) 44,46
Apollinaire, Guillaume [poet] (1880–1918) 109
Aquinas, Saint Thomas [theologian] (1225–1274) 260
Artaud, Antonin [artist / writer] (1896–1948) 260, 261, 292, 293
Augstine, Saint [theologian] (354–430) 260
Aurier, Albert [art critic] (1865–1892) 48, 55, 67, 71, 94, 283, 284

Baillarger, Jules [physician] (1809–1890) 175, 201, 203, 207
Balzac, de, Honoré [writer] (1799–1850) 53, 302
Bargue, Charles [artist] (d.1883) 14, 15, 32, 287
Baudelaire, Charles [poet] (1821–1867) 108, 300
Bayle, Antoine [physician] (1799–1858) 183, 201
Begemann, Margot [friend to Vincent van Gogh] (1841–1907) 26, 41
Bernard, Claude [physiologist] (1813–1878) 104
Bernard, Émile [artist] (1868–1941) 17, 44, 46, 47, 51, 52, 70, 71, 75, 84, 87, 94, 183, 205, 250–253, 284, 302
Bierre de Boismont, Alexandre [physician] (1797–1881) 203
Bing, Siegfried [art dealer] (1838–1905) 47
Bismarck, von, Otto [statesman] (1815–1898) 79, 81
Blanche, Antoine [physician] (fl. 1890) 95
Blanche, Émile [physician] (1820–1893) 95
Blanche, Jacques-Émile [artist] (1861–1942) 95
Boch, Anna [artist] (1848–1933) 283
Boch, Eugène [artist] (1855–1941) 49, 52, 96, 283
Bonger, Andries [brother-in-law to Theo van Gogh] (1861–1936) 46–48, 67, 70, 93–95, 146, 183, 184, 204
Boussod, Léon [art dealer] (1826–1893) 26, 44, 48, 50, 67, 70, 95
Boyle, Robert [chemist] (1627–1691) 207
Bredt, K. Julius [chemist] (1855–1937) 123
Breitner, George [artist] (1857–1923) 52
Breton, Jules [artist / writer] (1827–1906) 31, 39, 53
Brontë, Charlotte [writer] (1816–1855) 53
Browne, Thomas [philosopher] (1605–1682) 260
Brunschwig, Hieronymus [scientist] (c.1450–c.1512) 118
Bruyas, J.L. Alfred [art collector] (1821–1877) 56, 65
Bunyan, John [writer] (1628–1688) 53

Capuron, Joseph [physician] (1767–1850) 43, 211

Subject Index

326